Macromolecular Biorecognition
Principles and Methods

Experimental Biology and Medicine

Macromolecular Biorecognition: *Principles and Methods,* edited by *Irwin Chaiken, Emilia Chiancone, Angelo Fontana, and Paolo Neri,* 1987

Molecular Basis of Lymphokine Action, edited by *David R. Webb, Carl W. Pierce, and Stanley Cohen,* 1987

Biology of Copper Complexes, edited by *John R. J. Sorenson,* 1987

Epstein–Barr Virus and Human Disease, edited by *P. H. Levine D. V. Ablashi, M. Nonoyama, G. R. Pearson, and R. Glaser,* 1987

Methods of Protein Sequence Analysis • 1986, edited by *Kenneth A. Walsh,* 1987

Regulation of Immune Gene Expression, edited by *Marc Feldmann and Andrew McMichael,* 1986

Biological Methylation and Drug Design, edited by *Ronald T. Borchardt, Cyrus R. Creveling, and Per Magne Ueland,* 1986

Retroviruses and Human Pathology, edited by *Robert C. Gallo, Dominique Stehelin, and Oliviero E. Varnier,* 1985

Directed Drug Delivery, edited by *Ronald T. Borchardt, Arnold Repta, and Valentino Stella,* 1985

Immune Regulation, edited by *Marc Feldmann and N. A. Mitchison,* 1985

Human T Cell Clones, edited by *Marc Feldmann, Jonathan R. Lamb, and James N. Woody,* 1985

Inositol and Phosphoinositides, edited by *John E. Bleasdale, Joseph Eichberg, and George Hauser,* 1985

Growth, Cancer, and the Cell Cycle, edited by *Philip Skehan and Susan J. Friedman,* 1984

Ir Genes, edited by *Carl W. Pierce, Susan E. Cullen, Judith A. Kapp, Benjamin D. Schwartz, and Donald C. Shreffler,* 1983

Methods in Protein Sequence Analysis, edited by *Marshall Elzinga,* 1982

Inflammatory Diseases and Copper, edited by *John R. J. Sorenson,* 1982

Macromolecular Biorecognition

Principles and Methods

Edited by

Irwin Chaiken
National Institutes of Health, Bethesda

Emilia Chiancone
University of Rome "La Sapienza"

Angelo Fontana
University of Padua

and Paolo Neri
University of Siena

Humana Press • Clifton, New Jersey

Crescent Manor
PO Box 2148
Clifton, NJ 07015

Library of Congress Cataloging-in-Publication Data

Macromolecular biorecognition: principles and methods / edited by
Irwin Chaiken...[et al.].
p. cm.—(Experimental biology and medicine)
Based on the Conference on "Mechanisms of Recognition in Biological Macromolecules," held in Siena, Italy, Sept. 4–6, 1986 and co-sponsored by the National Research Council of Italy and others.
Includes indexes.
ISBN 0-89603-141-1
1. Biopolymers—Congresses. 2. Molecular association—Congresses. 3. Binding sites (Biochemistry)—Congresses. 4. Cellular recognition—Congresses. I. Chaiken, Irwin M. II. Conference on "Mechanisms of Recognition in Biological Macromolecules" (1986: Siena, Italy) III. Consiglio nazionale delle ricerche (Italy) IV. Series: Experimental biology and medicine (Clifton, NJ)
[DNLM: 1. Antigens—immunology—congresses. 2. Chromatography, Affinity—methods—congresses. 3. Macromolecular Systems—congresses. 4. Nucleic Acids—metabolism—congresses. 5. Proteins—metabolism—congresses. QU 55 M174 1986]
QP514.2.M34 1987
574.19'2—dc19 87-29304
CIP

Preface

Specific biomolecular interactions occur in practically every biological process, usually involving macromolecules such as proteins and nucleic acids. Thus, it is not surprising that great attention has been devoted by biologists, chemists, and physical chemists to the general phenomenon of macromolecular recognition in order to gain a better understanding of the principles, forces, and energetics that determine the often remarkable specificities of such interactions as those of enzymes and their substrates and inhibitors, antibodies with antigens, hormones with their receptors, and DNA with drugs. These studies have allowed the unraveling of some basic principles of molecular recognition, such as complementarity, specificity, and dynamics of the interacting species, and at the same time have led to development of analytical and separation techniques based on biological specificity (e.g., affinity chromatography and other bioaffinity methods).

This volume was assembled to address current research and developing ideas in biomolecular recognition through the contributions presented at the Conference on "Mechanisms of Recognition in Biological Macromolecules" held in Siena, Italy, September 4–6, 1986. The Conference was cosponsored by the National Research Council of Italy, the Italian Biochemical Society, the University of Siena, and the International Interest Group in Biorecognition Technology.

It was the intention of the organizers to bring together researchers with a broad range of scientific background in chemistry, biochemistry, biophysics, and immunology in order to discuss the general phenomenon of biorecognition in proteins and nucleic acids and the use of biorecognition principles to solve analytical and separation problems. The major areas chosen for discussion included: principles and methods for the design of recognition surfaces in proteins; specific interactions and functional regulation of proteins; interactions of nucleic acids; analytical and preparative bioaffinity methods; and immunological recognition and development of synthetic vaccines. Scientists from Italy and abroad attended the Conference. The intimate environment prompted a stimulating interdisciplinary forum among scientists working in different fields and looking at a common phenomenon from many specific viewpoints.

The editors hope that this volume will be useful to all researchers interested in understanding the general problem of interacting biomolecules and that it will stimulate new research ideas among its readers much as the meeting did among the conferees in Siena.

Irwin Chaiken
Emilia Chiancone
Angelo Fontana
Paolo Neri

Acknowledgments

The organizers would like to express their gratitude, for financial support of the Conference, to the University of Siena, the National Council of Research of Italy, Italian Society of Biochemistry, Comune di Siena, Regione Toscana, Ente Provinciale e Azienda Autonoma del Turismo di Siena, Monte dei Paschi di Siena, and Cassa Rurale ed Artigiana di Monteriggioni. We especially thank for their contributions the pharmaceutical industries Farmitalia-C. Erba, Fidia Research Laboratories, Merck Sharpe and Dohme Italia, Pfizer, Sclavo, and Sorin Biomedica, as well as the firms Applied Biosystems, Beckman Analytical, and Pharmacia.

The efficient cooperation of Conventur S. r. l., Siena, in the organization of this Conference is gratefully acknowledged. The participants in the Conference very much enjoyed, besides science itself, the scenic frame afforded by the beautiful city of Siena, including the festive atmosphere at the reception hosted in the City Hall and the dinners in the Contrada della Selva and in Piazza del Campo.

Contents

IV. Analytical and Preparative Bioaffinity Methods

Contributors

L. Adorini • *(279) Laboratory of Pathology, ENEA C. R. E. Casaccia, C. P. 2400, 00100 Rome (Italy)*

G. Alagona • *(13) Istituto di Chimica Quantistica ed Energetica Molecolare del CNR, Via Risorgimento 35, 56100 Pisa (Italy)*

A. Albertini • *(291) Institute of Chemistry, Faculty of Medicine, University of Brescia, Piazzale Ospedali Civili 1, 25100 Brescia (Italy)*

E. Alessio • *(199) Department of Chemical Sciences, University of Trieste, Piazzale Europa 1, 34127 Trieste (Italy)*

G. Amiconi • *(81, 101, 117) CNR Center of Molecular Biology, Department of Biochemical Sciences, University of Rome "La Sapienza," Piazzale A. Moro 5, 00185 Rome (Italy)*

A. G. Amit • *(219) Départmént d'Immunologie, Institut Pasteur 75724 Paris Cedex 15 (France)*

S. Ando • *(29) National Institute of Diabetes and Digestive and Kidney Diseases, National Institutes of Health, Bethesda, MD 20892 (USA)*

G. Antoni • *(251) Sclavo S. p. A. Research Center, Via Fiorentina 1, 53100 Siena (Italy)*

G. Antonini • *(131) Department of Experimental Medicine and Biochemical Sciences, University of Rome "Tor Vergata," Via O. Raimondo, 00173 Rome (Italy)*

E. Appella • *(279) Laboratory of Cell Biology, National Cancer Institute, NIH, Bethesda, Maryland 20892 (USA)*

P. Arosio • *(291) Dipartimento di Scienze e Technologie Biomediche, Via Olgettina 60, 20132 Milano (Italy)*

P. Ascenzi • *(81, 101, 117) CNR Center of Molecular Biology, Department of Biochemical Sciences, University of Rome "La Sapienza," Piazzale A. Moro 5, 00185 Rome (Italy)*

J. Ausio • *(153) Department of Polymer Research, The Weizmann Institute of Science, Rehovot 76100 (Israel)*

C. Balestrieri • *(213) Istituto di Chimica e Chimica Biologica, Via Costantinopoli 16, 80138 Napoli (Italy)*

T. BECCARI • *(309) Istituto di Chimica Biologica, Facoltá di Farmacia e Scienze, Via del Giochetto 37, 05100 Perugia (Italy)*

M. BOLOGNESI • *(81, 101, 117) Department of Genetics and Microbiology, Section of Crystallography, University of Pavia, Via Taramelli 16, 27100 Pavia (Italy)*

F. BONELLI • *(273) Eniricerche, Polypeptide Synthesis Department, 00015 Monterotondo (Italy)*

D. BORASCHI • *(251) Sclavo S. p. A. Research Center, Via Fiorentina 1, 53100 Siena (Italy)*

G. BORIN • *(147) Department of Organic Chemistry, Biopolymer Research Center, University of Padua, Via Marzolo 1, 35131 Padua (Italy)*

C. BOVE • *(279) Laboratory of Pathology, ENEA C. R. E. Casaccia, 00100 Rome (Italy)*

L. BRACCI • *(259) Dipartimento di Biologia Evolutiva, Universitá di Siena, 53100 Siena (Italy)*

M. BRUNORI • *(127) Department of Biochemical Sciences and CNR Center of Molecular Biology, University of Rome "La Sapienza," Piazzale A. Moro 5, 00185 Rome (Italy)*

M. CACACE • *(341) CNR Institute of Protein Chemistry and Enzymology, Via Toiano 6, 80072 Arco Felice, Naples (Italy)*

S. CAUCI • *(199) Department of Biochemistry, Biophysics and Macromolecular Chemistry, University of Trieste, Piazzale Europa 1, 34127 Trieste (Italy)*

S. CENSINI • *(251) Sclavo S. p. A. Research Center, Via Fiorentina 1, 53100 Siena (Italy)*

I. M. CHAIKEN • *(29, 235) National Institute of Diabetes and Digestive and Kidney Diseases, National Institutes of Health, Bethesda, MD 20892 (USA)*

E. CHIANCONE • *(321) CNR Center of Molecular Biology, Department of Biochemical Sciences, University of Rome "La Sapienza," Piazzale A. Moro 5, 00185 Rome (Italy)*

F. CHIOLERIO • *(303) Societá Prodotti Antibiotici, Via Bielle 8, 20123 Milano (Italy)*

G. COLONNA • *(143) Istituto di Chimica e Chimica Biologica, I Facoltá di Medicina e Chirurgia, Via Costantinopoli 16, 80138 Napoli (Italy)*

G. CORRADIN • *(273) Institute of Biochemistry, University of Lausanne, Ch. des Boveresses, 1066 Epalinges (Switzerland)*

B. CURTI • *(333) Dipartimento di Fisiologia e Biochimica Generale, Sezione Chimica Biologica, Universitá di Milano, Via Celoria 26, 20133 Milano (Italy)*

C. DALZOPPO • *(51, 235) Dipartimento di Chimica Organica, Universitá di Padova, Via Marzolo 1, 35131 Padova (Italy)*

M. DARSLEY • *(279) Laboratory of Cell Biology, National Cancer Institute, NIH, Bethesda, MD 20892 (USA)*

G. DEL GIUDICE • *(273) WHO Immunology Research and Training Center, University of Geneva, 1211 Geneva 4 (Switzerland)*

P. DE SANTIS • *(169) Dipartimento di Chimica, Universitá di Roma "La Sapienza," Piazzale A. Moro 5, 00185 Rome (Italy)*

A. DI TOMMASO • *(259) CRISMA, Universitá di Siena, Policlinico Le Scotte, 53100 Siena (Italy)*

G. DORIA • *(279) Laboratory of Pathology, ENEA C. R. E. Casaccia, 00100 Rome (Italy)*

H. EISENBERG • *(153) Department of Polymer Research, The Weizmann Institute of Science, Rehovot 76100 (Israel)*

C. EMILIANI • *(309) Istituto di Chimica Biologica, Facoltá di Farmacia e Scienze, Via del Giochetto 37, 06100 Perugia (Italy)*

H. D. ENGERS • *(273) WHO Immunology Research and Training Center, University of Geneva, 1211 Geneva 4 (Switzerland)*

G. FASSINA • *(29) National Institute of Diabetes and Digestive and Kidney Diseases, National Institutes of Health, Bethesda, MD 20892 (USA)*

B. FILIPPI • *(147) Department of Organic Chemistry, Biopolymer Research Center, Via Marzolo 1, 35131 Padua (Italy)*

E. FILIPPINI • *(303) Societá Prodotti Antibiotici, Via Biella 8, 20123 Milano (Italy)*

A. FONTANA • *(51, 235) Dipartimento di Chimica Organica, Universitá di Padova, Via Marzolo 1, 35131 Padova (Italy)*

G. L. FOSSATI • *(333) Recordati S. p. A., Via M. Civitali 1, 20148 Milano (Italy)*

P. FUSI • *(333) Dipartimento di Fisiologia e Biochimica Generali, Universitá di Milano, Via Celoria 26, 20133 Milano (Italy)*

M. GATTONI • *(321) CNR Center of Molecular Biology, Department of Biochemical Sciences, University of Rome "La Sapienza," Piazzale A. Moro 5, 00185 Rome (Italy)*

P. GHIARA • *(251) Sclavo S. p. A. Research Center, Via Fiorentina 1, 53100 Siena (Italy)*

C. GHIO • *(13) Istituto di Chimica Quantistica ed Energetica Molecolare del CNR, Via Risorgimento 35, 56100 Pisa (Italy)*

W. A. GIBBONS • *(259) Dept. of Pharmaceutical Chemistry, School of Pharmacy, University of London, 29 Brunswick Sq., WC1N 1AX London (UK)*

A. Giovane • *(213) Instituto di Chimica e Chimica Biologica, Via Costantinopoli 16, 80138 Napoli (Italy)*E. Grazi • *(139) Istituto di Chimica Biologica, Universitá di Ferrara, Via Borsari 46, 44100 Ferrara (Italy)*

K. O. Greulich • *(153) Department of Polymer Research, The Weizmann Institute of Science, Rehovot 76100 (Israel)*

M. Guarneri • *(81, 101, 117) Department of Pharmaceutical Sciences, University of Ferrara, Via Scandiana 21, 44100 Ferrara (Italy)*

M. Harper • *(219) Départment d'Immunologie, Institut Pasteur, 75724 Paris Cedex 15 (France)*

P. A. Kollman • *(13) Department of Pharmaceutical Chemistry, S-296, University of California, San Francisco, CA 94143 (USA)*

C. Iacobello • *(291) Institute of Chemistry, Faculty of Medicine, University of Brescia, Piazzale Ospedali Civili 1, 25100 Brescia (Italy)*

P.-H. Lambert • *(273) WHO Immunology Research and Training Center, University of Geneva, 1211 Geneva 4 (Switzerland)*

A. La Pegna • *(143) Istituto di Chimica e Chimica Biologica, I Facoltá di Medicina e Chirurgia, Via Costantinopoli 16, 80138 Napoli (Italy)*

F. Lema • *(219) Départment d'Immunologie, Institut Pasteur, 75724 Paris Cedex 15 (France)*

L. Leoni • *(169) Centro di Studio per gli Acidi Nucleici, Dipartimento di Genetica e Biologia Molecolare, Universitá di Roma "La Sapienza," Piazzale A. Moro 5, 00185 Rome (Italy)*

X. Liang • *(29) Department of Chemistry, Nankai University, Tianjin, Peoples Republic of China*

L. Lozzi • *(259) CRISMA, Universitá di Siena, Policlinico Le Scotte, 53100 Siena (Italy)*

S. Magnaghi • *(303) Societá Prodotti Antibiotici, Via Biella 8, 20123 Milano (Italy)*

F. Malatesta • *(131) Department of Experimental Medicine and Biochemical Sciences, University of Rome "Tor Vergata," Via O. Raimondo, 00173 Rome (Italy)*

M. Malcovati • *(303) Dipartimento di Biologia e Genetica per le Scienze Mediche, Universitá di Milano, Via Viotti 5, 20133 Milano (Italy)*

F. Marchiori • *(147) Department of Organic Chemistry, Biopolymer Research Center, University of Padua, Via Marzolo 1, 35131 Padua (Italy)*

S. Marciani Magno • *(185) Dipartimento di Scienze Farmaceutiche, Universitá di Padova, Via Marzolo 5, 35131 Padova (Italy)*

R. A. Mariuzza • *(219) Départment d'Immunologie, Institut Pasteur, 75724 Paris Cedex 15 (France)*

P. MASCAGNI • *(259) Department of Pharmaceutical Chemistry, School of Pharmacy, University of London, 29 Brunswick Sq., WC1N 1AX London (UK)*

E. MENEGATTI • *(81, 101, 117) Department of Pharmaceutical Sciences, University of Ferrara, Via Scandiana 21, 44100 Ferrara (Italy)*

G. MESTRONI • *(199) Department of Chemical Sciences, Universtity of Trieste, Piazzale Europa 1, 34127 Trieste (Italy)*

H. L. MONACO • *(69) Centro di Studi sui Biopolimeri, Istituto di Chimica Organica, Universitá di Padova, Via Marzolo 1, 35131 Padova (Italy)*

E. MONTI • *(333) Dipartimento di Fisiologia e Biochimica Generale, Sezione Chimica Biologica, Universitá di Milano, Via Celoria 26 20133 Milano (Italy)*

L. NENCIONI • *(251) Sclavo S. p. A. Research Center, Via Fiorentina 1, 53100 Siena (Italy)*

P. NERI • *(259) CRISMA, Universitá di Siena, Policlinico Le Scotte, 53100 Siena (Italy)*

N. NICCOLAI • *(259) Dipartimento di Chimica, Universitá di Siena, 53100 Siena (Italy)*

A. ORLACCHIO • *(309) Istituto di Chimica Biologica, Facoltá di Farmacia e Scienze, Via del Giochetto 37, 06100 Perugia (Italy)*

C. PALERMO • *(169) Centro di Studio per gli Acidi Nucleici del CNR, Dipartimento di Genetica e Biologia Molecolare, Universitá di Roma "La Sapienza," Piazzale A. Moro 5, 00185, Roma (Italy)*

A. PALLESCHI • *(169) Dipartimento di Chimica, Universitá di Roma "La Sapienza," Piazzale A. Moro 5, 00185 Rome (Itlay)*

G. PALÚ • *(185) Istituto di Microbiologia, Universitá di Padova, via Gabelli 63, 35100 Padova (Italy)*

M. PALUMBO • *(185) Dipartimento di Chimica Organica, Universitá di Padova, Via Marzolo 1, 35131 Padova (Italy)*

F. PERIN • *(251) Sclavo S. p. A. Research Center, Via Fiorentina 1, 53100 Monterotondo (Italy)*

A. PESSI • *(273) Eniricerche, Polypeptide Synthesis Department, 00015 Monterotondo (Italy)*

S. PETRENI • *(259) CRISMA, Universitá di Siena, Policlinico Le Scotte, 53100 Siena (Italy)*

S. E. V. PHILLIPS • *(219) Astbury Department of Biophysics, University of Leeds, Leeds LS2 9JT (UK)*

R. J. POLJAK • *(219) Départment d'Immunologie, Institut Pasteur, 75724 Paris Cedex 15 (France)*

R. Presentini • *(251) Sclavo S. p. A. Research Center, Via Fiorentina 1, 53100 Siena (Italy)*

D. Pulcini • *(143) Istituto di Chimica e Chimica Biologica, I Facoltá di Medicina e Chirurgia, Via Costantinopoli 16, 80138 Napoli (Italy)*

F. Quadrifoglio • *(199) Department of Biochemistry, Biophysics and Macromolecular Chemistry, University of Trieste, Piazzale Europa 1, 34127 Trieste (Italy)*

L. Quagliuolo • *(213) Istituto di Chimica e Chimica Biologica, Via Costantinopoli 16, 80138 Napoli (Italy)*

R. Ragone • *(143) Istituto di Chimica e Chimica Biologica, I Facoltá di Medicina e Chirugia, Via Costantinopoli 16, 80138 Napoli (Italy)*

C. Roscigno • *(143) Istituto di Chimica e Chimica Biologica, I Facoltá di Medicina e Chirurgia, Via Costantinopoli 16, 80138 Napoli (Italy)*

G. Ruggeri • *(291) Institute of Chemistry, Faculty of Medicine, University of Brescia, Piazzale Ospedali Civili 1, 25100 Brescia (Italy)*

M. Rustici • *(259) CRISMA, Universitá di Siena, Policlinico Le Scotte, 53100 Brescia (Italy)*

A. Sada • *(341) CNR Institute of Protein Biochemistry and Enzymology, Via Toiano 6, 80072 Arco Felice, Naples (Italy)*

A. Santucci • *(259) CRISMA, Universitá di Siena, Policlinico Le Scotte, 53100 Siena (Italy)*

P. Sarti • *(131) CNR Center of Molecular Biology, Department of Biochemical Sciences, University of Rome "La Sapienza," Piazzale A. Moro 5, 00185 Rome (Italy)*

M. Savino • *(169) Centro di Studio per gli Acidi Nucleici del CNR e Dipartimento di Genetica e Biologia Molecolare, Universitá di Roma "La Sapienza," Piazzale A. Moro 5, 00185 Rome (Itlay)*

R. Sciarra • *(309) Istituto di Chimica Biologica, Facoltá di Farmacia e Scienze, Via del Giochetto, 37, 06100 Perugia (Italy)*

D. Seger • *(159) Department of Polymer Research, The Weizmann Institute of Science, Rehovot 76100 (Israel)*

L. Servillo • *(213) Istituto di Chimica e Chimica Biologica, Via Constantinopoli 16, 80138 Napoli (Italy)*

Y. Shai • *(29) National Institute of Diabetes and Digestive and Kidney Diseases, National Institutes of Health, Bethesda, MD 20892 (USA)*

G. Siligardi • *(259) Department of Pharmaceutical Chemistry, University of London, 29 Brunswick Sq., WC1N 1AX, London (UK)*

P. Soldani • *(259) CRISMA, Universitá di Siena, Policlinico Le Scotte, 53100 Siena (Italy)*

P. Soldani • *(259) CRISMA, Universitá di Siena, Policlinico Le Scotte, 53100 Siena (Italy)*

P. Spadon • *(69) Centro di Studi sui Biopolimeri, Istituto di Chimica Organica, Universitá di Padova, Via Marzolo 1, 35131 Padova (Italy)*

J. L. Stirling • *(309) Department of Biochemistry, King's College, University of London, Atkins Buildings, Campden Hill, London W8 7AH (UK)*

P. Stiuso • *(143) Istituto di Chimica e Chimica Biologica, I Facoltá di Medicina e Chirurgia, Via Costantinopoli 16, 80138 Napoli (Italy)*

A. Tagliabue • *(251) Sclavo S. p. A. Research Center, Via Fiorentina 1, 53100 Siena (Italy)*

A. R. Togna • *(273) Eniricerche, Polypeptide Synthesis Department, 00015 Monterotondo (Italy)*

G. Valentini • *(303) Dipartimento di Biologia e Genetica per le Scienze Mediche, Universitá di Milano, Via Viotti 5, 20133 Milano (Italy)*

M. H. V. Van Regenmortel • *(227) Institut de Biologie Moléculaire et Cellulaire CNRS, 15 Rue Descartes, 67084 Strasbourg Cédex (France)*

P. Vecchini • *(131) Department of Biochemical Sciences and CNR Center of Molecular Biology, University of Rome "La Sapienza," Piazzale A. Moro 5, 00185 Rome (Italy)*

A. Verdini • *(273) Eniricerche, Polypeptide Synthesis Department, 00015 Monterotondo (Italy)*

L. Villa • *(251) Sclavo S. p. A. Research Center, Via Fiorentina 1, 53100 Siena (Italy)*

C. Vita • *(51, 235) Dipartimento di Chimica Organica, Universitá di Padova, Via Marzolo 1, 35131 Padova (Italy)*

G. Volpini • *(251) Sclavo S. p. A. Research Center, Via Fiorentina 1, 53100 Siena (Italy)*

E. Wachtel • *(153) Department of Polymer Research, The Weizmann Institute of Science, Rehovot 76100 (Israel)*

M. T. Wilson • *(131) Department of Chemistry, University of Essex, Colchester C04 3SQ, Essex (UK)*

S. Wodak • *(3) Université Libre de Bruxelles, Unité de Conformation des Macromolécules Biologiques, C. P. 160, Avenue Paul Héger P2, 1050 Brussels (Belgium)*

G. Zanotti • *(69) Centro di Studi sui Biopolimeri, Dipartimento di Chimica Organica, Universitá di Padova, Via Marzolo 1, 35131 Padova (Italy)*

SECTION I

SPECIFIC INTERACTIONS IN PROTEINS: MOLECULAR ASPECTS AND FUNCTIONAL REGULATION

COMPUTER GRAPHICS AS AN AID IN PROTEIN STRUCTURE-FUNCTION ANALYSIS AND DESIGN.

Shoshana J. Wodak

Université Libre de Bruxelles, U.C.M.B. CP160

av. P. Héger P2, 1050 Brussels, Belgium

1. INTRODUCTION

The first application of computer graphics to the study of molecules was made in the mid-sixties. Cyrus Levinthal and Robert Langridge, then at M.I.T., were interested in the M.A.C. (Multiple Access Computer) project, an ancestor of present day C.A.D./C.A.M. systems. It occurred to them that a computer driven cathode ray tube, a very preliminary version of presently available graphic systems, could be a valuable tool in molecular biology.

These pioneers had an extremely ambitious objective: to simulate on the computer the physical forces which control formation of the 3D structure of proteins, starting from the amino-acid sequence. Computer generated image was to play an important role in this enterprise, providing the indispensible visual contact between the researcher and the concepts embedded in the complex 3D models of macromolecules.

Today, 20 years later, the original objective has still not been reached, but research in this field has been both extensive and productive, leading to the development of a number of new disciplines of which conformational analysis of biological molecules and molecular graphics are leading examples.

With recent progress in performance of computer hardware and software, but even more, with recent advances in genetic engineering techniques, the prospects for successful applications of molecular modelling to proteins has markedly increased.

The present potential for de novo design of proteins is still limited due to our incomplete understanding of the factors which determine the relationship between the covalent structure of the polypeptide chain (the amino-acid sequence) and the 3D structure, the stability and functional properties of the folded macromolecule. It has nonetheless been possible to design a small functional RNA-binding protein (1), and attempts are on the way in a number of laboratories to obtain small proteins containing simple repeating sequence motifs by de novo design and gene synthesis (2,3).

The mounting of protein structural changes on a basal skeleton is a much more attractive possibility. The potential of the strategy is naturally explicit in the diversity of immunological responses achieved by in-vivo recombination of DNA segments and somatic mutation. It is also clearly, albeit less dramatically, evident from the common structural elements which are being found in proteins which display very different functions and little or no sequence homology.

The production of mutants where one or a very limited number of amino-acids are replaced in a protein is already a very active field of endeavour which is bound to grow rapidly (3-7). The main emphasis is on definition and modulation of specific functions and it is the longterm hope that it will be possible to produce proteins with predetermined new functions. For systematic protein engineering at this level, the three-dimensional structure at appropriate high resolution is required. At present about 100 protein structures are known and information on them is available in a data bank compiled in Brookhaven, U.S.A.

Improvements in crystallization techniques as well as bright, tunable X-ray sources and more efficient data collection devices (area detectors) have increased the ability for determining additional structures.

Once structural information is available, particularly for protein-ligand (where the ligand can be DNA, RNA, or another protein) or protein-substrate analog complexes, then it is possible to try to improve, say, substrate affinity by optimizing the local interactions between the bound substrate and the protein or try to improve the reaction rate by changing the catalytic site so that reaction intermediates bind better.

The design of experiments of this type requires

accurate model building which incorporates sophisticated techniques of molecular graphics and numerical simulations. There are bound to be unsuspected subtleties of protein structure and intricate consequences of side chain substitutions which confound simple predictions. It is the longterm hope that these techniques will ultimately constitute genuine Computer Aided Design systems for proteins and that their use will eliminate a number of tedious experimental steps.

2. THE COMPONENTS OF C.A.D. SYSTEMS FOR PROTEINS.

The widely used C.A.D. systems in civil and electronic engineering contain two major components: visual representation of the object to be designed and numerical simulations of its functional properties.

In the case discussed here the object is a protein, something we are not used to manipulate on an everyday basis, but otherwise the principles are the same. While visual representation and manipulation on a cathode ray tube of complex molecular models has reached an appreciable degree of sophistication, much progress remains to be made before the physical and chemical properties of displayed models can be reliably simulated.

2.1 Molecular Graphics.

Extracting the wealth of information obtained by protein crystallography has always been a challenging problem.

In the past, scientists were building models out of wire and plastic. Such models were hard to make, difficult to examine and liable to damage.

Present day graphic facilities offer a variety of ways for displaying protein structure in color: bond connectivity, surfaces, volumes and physical-chemical properties of amino-acid side-chains (8-10). Calligraphic (vector) displays offer the highest image resolution and fastest capacity for real-time manipulation of molecular models. Raster systems based on more standard TV technology are undergoing rapid improvements which are beginning to be competitive. Real-time manipulation of models is the computerized substitute for manual tinkering. The computer can move and rotate the displayed picture or modify the model according to the precise instructions given by the user through a variety of control devices such as light pens, dials and data tablets.

With interactive computer graphics, models displayed and coordinate data are rigorously related. It is therefore possible to combine quantitative measures of geometric parameters or physical properties with visual impressions interactively.

2.2 Simulation Methods

The physical properties of protein models are evaluated by the spatial coordinates of their atoms. The forces and interaction energies between atoms or chemical groups within biological macromolecules are represented using empirical force fields, with parameters derived from the analysis of known structures, mostly small organic molecules (11).

Typically the energy expressions contain contributions which simulate chemical bond stretching and bond angle bending, and also Van der Waals forces, repulsive at very short distances and attractive at distances corresponding to normal contacts between non-bonded atoms. The representation of hydrogen bonds, of interactions with the solvent and of other electrostatic interactions raises more difficult problems such as evaluating the dielectric and hydrophobicity effects. In most computations the effect of surrounding water molecules has been ignored, although this is rapidly changing with increasing access to supercomputers. Energy minimization is most commonly used to find "stable" conformations of proteins (12). When the minimization starts from atomic positions defined by high resolution X-ray crystallography, an energy minimum (a local one) is usually found near the crystallographic structure. Further away from the crystal structure, there may be other local minima in the highly complex energy surface, which remain unexplored. Some of these might possibly represent conformations taken by the protein in solution. In addition to finding energy minima, energy calculations can be used to simulate the dynamic behaviour of the protein which, we know, must also play a crucial role in function.

An appropriate description of atomic movements near an energy minimum (thermal vibrations) or from one minimum to another is provided by Molecular Dynamics simulations (13) where calculated forces are used to solve Newton's equation of motion at very short time intervals. Groups in Harvard,

Argonne National Labs. and Groningen have pioneered the application of molecular dynamics to biological macromolecules about ten years ago. It has become a fairly standard technique since. It is particularly useful in exploring conformations around the crystal structure, because unlike minimization, it is able to overcome energy barriers through thermal vibrations.

However, only very fast movements may be simulated by the technique as it stands today. Typically, Molecular Dynamics simulations of protein internal motions deal with ranges of 10 to 200 psec (1 psec = 10^{-12} sec). These very fast movements are localized and have no appreciable enthalpy of activation. Concerted movements of large parts of the protein structure like those which are associated with ligand binding or other functional properties occur on much longer time scales.

The procedures whereby large conformational changes are studied today involve choosing _a priori_ a reaction pathway and minimizing the energy or simulating the dynamic behavior along this pathway.

Similarly, the evaluation of Free Energies as opposed to Enthalpies, also require lengthy molecular dynamics calculations and can only be performed for very simple cases. _A fortiori_, simulating the entire protein folding process is presently impossible.

3. OBJECTIVES WITHIN REACH OF TODAY'S COMPUTER MODELING AND SIMULATION METHODS.

The present state of knowledge precludes the prediction of three-dimensional structure or even the prediction of accurate secondary structure from the amino acid sequence. This is because interactions between regions which lie apart on the polypeptide chain play important roles in establishing secondary end tertiary structures, but such interactions are not properly taken into account in present day methods.

Progress in prediction could result from simulation of the folding process, but this is very unlikely in light of the arguments given above. Alternatively it would be achieved from developing a computerized data base which would include structural and sequence data combined with higher level descriptions of folding motifs, topological features, and side-chain physical chemistry. The latter approach is already actively pursued in a number of European laboratories, such as Birckbeck College, London, in Uppsala and Brussels, as well as the U.S. In the meanwhile, complete prediction of tertiary structure is only possible when the structure of an homologous protein is available. There the structure is constructed by model building using graphic display and conformation energy calculations. The quality of the resulting structure will depend on the degree of sequence homology between the established and unknown structures, and recent work by Chothia and coll. (14) indicates that valid results may be expected in situations where this homology is about 50%. Other less ambitious but nonetheless difficult objectives are closer to being attained. For example, current computational procedures may be used to predict the conformation of a small organic molecule when bound to a protein (receptor or enzyme) whose 3D structure is known (15-16). This ability is extensively exploited to derive the structure of reaction intermediates of an enzymatic process starting from the crystal structure of an enzyme-inhibitor complex, or alternatively to design substrate analogues for therapeutic applications (drug design).

It is also possible to model small scale conformational changes in a protein of known structure. Scientists at Harvard (17) have correctly predicted the structural perturbations in Hemagglutinin from influenza virus due to a single amino-acid substitution (gly 146 --> asp). In Brussels we have been able to simulate the conformational transition between two monomers of insulin in the 2 Zn crystal (18) and to compute the structural perturbations in Deoxy Hemoglobin induced by the covalent attachment of Glutathione, a flexible tripeptide, at a Cystein residue in position 93 of the beta chain (19).

However, reliable modelling of the effect of single amino-acid replacements in proteins of known structure is still a problem. Due to the limitations mentioned above, only small structural perturbations can be simulated starting from the known crystal conformation of the protein. Simulation of larger structural changes such as those which may be introduced: by simultaneous substitution of many residues, by deletions or additions of entire segments of the polypeptide chain, or even computing perturbations introduced in flexible loops of the proteins which are in contact with solvent, is presently still very preliminary. It is hoped that Molecular Dynamics simulations in presence of surrounding water molecules will be useful to tackle these larger changes, but so far this approach has not been tested.

A promising avenue for modelling large structural variations in loops and other regions could come from an entirely different approach. From the analysis of the one hundred or so protein structures known today it appears that segments of the polypeptide chain with different sequences can adopt similar three-dimensional folds, and that observed folding patterns fall into a small number of topologically distinct families (20). One may imagine having a library of commonly observed "standard" structural motifs and fetching from it the one which best fits the sequence and structural data under consideration. Energy minimization or other numerical simulation procedures could then be used to obtain fine adjustments.

Besides evaluating the effects of "mutations" on structure it should be most useful to determine the resulting changes in physical and functional properties. Present computational tools are not powerful enough to allow for quantitative evaluation of changes in protein stability, in association constants with ligands or in pKas of relevant residues. But this situation may change in the near future.

4. PROTEIN C.A.D. SYSTEMS OF TOMORROW.

Systematic concertation of experimental and theoretical approaches will be the major motor for progress in protein engineering as a whole, and in model building methods in particular. It will help adjust simulation parameters and generate new insights, because, for the very first time in the history of this young science, we have the possibility to actually tinker with protein structures and probe our understanding.

Another important factor will be progress in computer technology. Availability of computers several orders of magnitude faster than the machines we have today will allow the performance of simulations ranging over much longer time scales. One will be able to consider larger systems in their solvent environment, evaluate free energies and consequently, association constants and kinetic parameters for a number of important phenomena. Faster computers will most of all provide the opportunity to improve the methods for simulating biologically relevant processes. Molecular graphics will be another direct beneficiary of advances in computer technology.

It will be the "humanized front end" of protein C.A.D. systems. All current display capabilities and more will be available at much lower cost in raster systems. Model manipulation will be performed through a much larger variety of feedback sensory control devices. Conformational energies and other numerical properties will be computed and minimized interactively for large and flexible systems. Future protein C.A.D. systems will be closely linked to data bases containing higher level descriptions of known protein structures as well as data on amino-acid and nucleic-acid sequences and on properties relevant to function.

REFERENCES

1. Gutte B., Daumigen, M. & Wittschiber (1979), Nature 281, 650-655.
2. Degrado, W.P., Wilcox, W. & Eisenberg, A. (1985), UCLA Symposia on Molecular and Cellular Biology, J. Cell. Biochem. Supplement 9B, 131.
3. UCLA Symposia on Molecular and Cellular Biology (1985), J. Cell. Biochem. Supplement 9B, 91-99.
4. Winter et.al., (1982), Nature, London 299, 756-758.
5. Fersht, A.R. et.al., (1985), Nature 314, 235-238.
6. Thomas, P.G., Russel, A.J. & Fersht, A.R., (1985), Nature 318, 375-376.
7. Craik, C.S. et.al., (1985), Science 228, 291-297.
8. Langridge, R., Ferrin, T.E., Kuntz, I.D. & Connolly, M.L., (1981), Science 211, 661-666.
9. Getzoff, E.D. et.al., (1985), Nature 306, 287-290.
10. Lesk, A.M., (1984), Trends in Biochemical Sciences V-VII.
11. Hagler, A.T., Huler, E. & Lifson, S., (1977), J. Am. Chem. Soc. 96, 5319-5326.
12. Nemethy, G. & Sheraga, H.A., (1977), Quarterly Review of Biophysics 10, 239-352.
13. Karplus, M. & McCammon, J.A., (1986), Scientific American 254, 30-39.
14. Chothia, C. & Lesk, A., (1986), EMBO Journal 5, 823.
15. Wipff, G. et.al., (1983), J. Am. Chem. Soc. 105, 997-1005.
16. Hangauer, D.G., Monzingo, A.F. & Matthews, B.W., (1985) Biochemistry 23, 5730-5741.
17. Shih, H.L., Brady, J. & Karplus, M., (1985), Proc. Natl Acad. Sci. USA 82, 1697-1700.
18. Wodak, S.J. et.al., (1985), J. Mol. Biol. 181, 317-322.

19. Wodak *et.al.*, (1986), J. Biol. Chem. (in press).
20. Jones, T.A. & Thirup, S., (1986), EMBO Journal 5, 819-822.

COMPUTATIONAL APPROACHES TO THE STUDY OF PROTEIN - LIGAND INTERACTIONS

Giuliano Alagona, Caterina Ghio and Peter A. Kollman°

C.N.R. Istituto di Chimica Quantistica ed Energetica Molecolare, 56100 Pisa (Italy)

°Department of Pharmaceutical Chemistry, University of California, San Francisco, CA 94143 USA

ABSTRACT

A model for the dihydroxyacetone phosphate ⟶glyceraldehyde phosphate reversible isomerization catalyzed by triose phosphate isomerase (TIM) has been subjected to various *ab initio* and molecular mechanical calculations coupled with computer graphics. Each of these techniques alone would give limited information on the enzyme catalyzed reaction mechanisms. The three methods combined allow us to get a useful picture of the way the enzyme exerts its catalytic effect. In addition it has been possible to rationalize and predict the effect of genetic mutations on some of the key active site residues in TIM. With the increasing applicability of genetic engineering techniques, a theoretical approach to predicting the effect of site-specific mutagenesis should be useful.

INTRODUCTION

The isomerization of the dihydroxyacetone phosphate (DHAP) to glyceraldehyde-3-phosphate (GAP), catalyzed by triosephosphate isomerase (TIM), should be regarded as a model to illustrate the use of computational techniques to study complex interactions, in addition to the importance of this mechanism "per se". In this context, we will devote ourselves more to the methodological aspects of the research than to the results, thoroughly examined in the previous papers (1,2), and will resort to actual calculations only to show how the method works. Obviously, the procedure employed is specific for the given system. Modifications should be made to study different systems, involving, e.g., charge separation or heavier atoms with complex electronic structures.

We do not claim to be the first to study an enzymatic reaction - previous calculations are known (for an extensive bibliography see Ref. 1) - but to our knowledge this was the first time that *ab initio* SCF, correlation energy and environmental effect calculations (3) have been carried out on all the chemical steps of an enzymatic reaction, along with molecular mechanical simulation of some steps.

For more recent studies on enzyme catalyzed reactions see: Warshel and Sussman (4), that simulate the effect of site-directed mutagenesis; Weiner et al. (5), that include into the calculations the entire enzyme and 200 water molecules; and the references quoted therein.

METHODS

The quantum mechanical calculations have been carried out (unless otherwise stated) on hydroxyacetone, HA, which is a model for DHAP, and on formate, which is a model of Glutammic acid, Glu 165 in TIM. Their geometries as well as those of the other species (intermediates and products) have been optimized at the 4-31G level.

First we determined with *ab initio* quantum mechanical methods the reaction mechanism and the barrier heights *in vacuo* for the various postulated steps of the reaction (the intermolecular and intramolecular proton transfers).

Then we tried to incorporate the effect of the environment into the calculations introducing a polarizing electrophile (NH_4^+) to mimic lysine 13.

In the meantime, we had model-built with the use of computer graphics three different structures of the enzyme that refined with the molecular mechanical minimization to somewhat different structures having nonetheless histidine 95, Lys 13 and Glu 97 placed in such a way as to most influence the reaction. We thus repeated the quantum mechanical calculations in the presence of the charges of these side chains. These charges appear as perturbations in the one-electron Hamiltonian. We introduced into the calculations also the charges of a mutant (Gln 95 instead of His 95), but in the same position as His. Then we also refined the coordinates of the mutant enzyme. We will describe in more detail the methods employed during the discussion of the results. The programs used are listed in the appendix.

AB INITIO REACTION MECHANISM

To study the various steps of the enzymatic reaction we resorted to simplified models of the partners that, due to their size, allowed us to go somewhat beyond the SCF level, with the introduction of MP2 correlation corrections (6). Once we determined which kind of improvement in the results the higher level of accuracy might produce, we studied the reaction mechanism that can be summarized as follows:

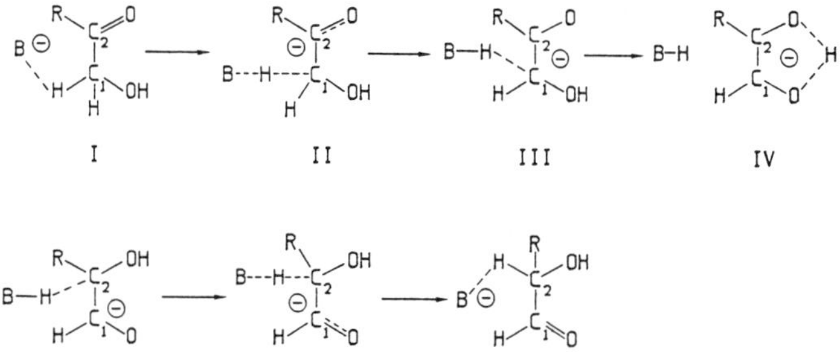

I II III IV

V VI VII

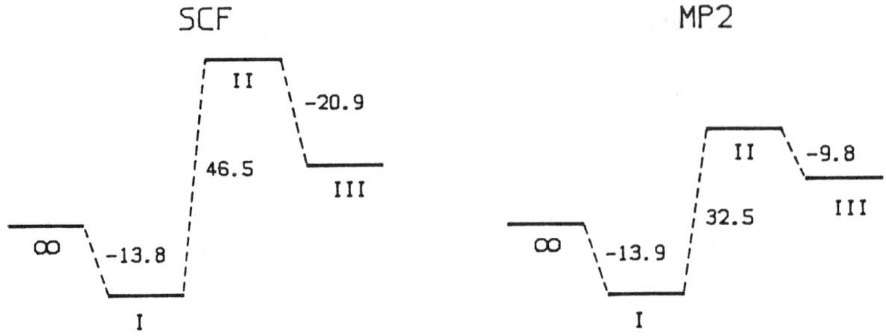

Fig. 1. Relative energies of the model with R = H at the SCF and MP2 levels for I, II and III (4-31G basis set), at fixed $R(C_1...O)$ and $R(C_1...H)$.

Since steps V-VII are essentially the reverse of steps I-III, that lead to the enediolate, and the intramolecular proton transfer (IV) occurs with a low barrier even *in vacuo* (1), let us focus on the first three steps. The relative energies of 4-31G model (R=H) at the SCF and MP2 levels for I, II and III are reported in Fig. 1. The substitution of the methyl group (Fig. 2) does not affect the results much and we may infer that this is true also at the MP2 level. These results have been obtained at fixed proton transfer distance ($R(C_1...O)=3.415$ Å and $R(C_1...H) = 1.715$ Å) determined with the use of models (1). Subse-

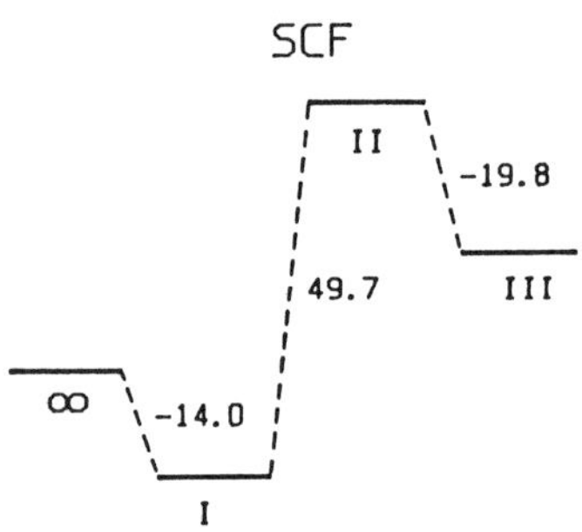

Fig. 2 Relative energies of the model with $R=CH_3$ at the SCF level for I, II and III (4-31G basis set) at fixed $R(C_1...O)$ and $R(C_1...H)$.

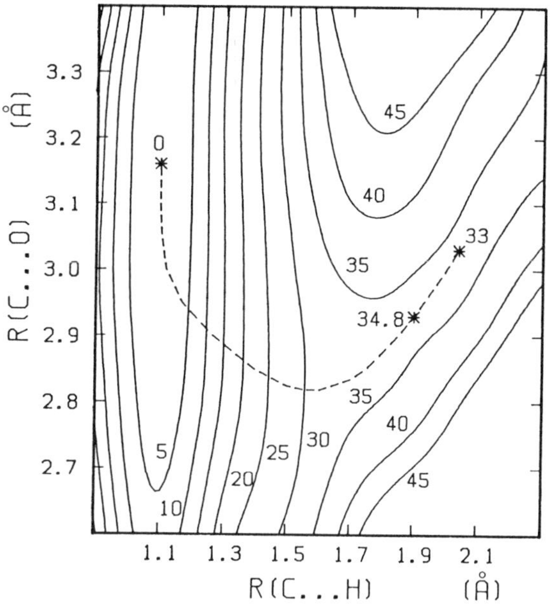

Fig. 3. Potential energy hypersurface at the 4-31G level in vacuo, for the conversion I to III ($R=CH_3$) as a function of the distance of the proton from the carbon atom and of the total distance between the C and O atoms. The stars stand for the reactants (I), the saddle point (II) and the enediolate intermediate (III). The dashed line stands for the probable reaction path. Note that, even though it represents the minimum energy pathway, it is not always perpendicular to the energy contours, and thus it is just a reasonable guess for the reaction coordinate. Energy contours in kcal/mol, distances in Ångstroms.

quently (2), we studied the potential energy hypersurface (Fig. 3) for this step of the reaction mechanism that is the most sensitive to the environment because it was completely inconsistent with a facile reaction. The saddle point on the hypersurface occurs at $R(C_1...O) = 2.93$ Å and $R(C_1...H) = 1.9$ Å. The energies for I, II and III with respect to the isolated reactants are reported in Fig. 4.

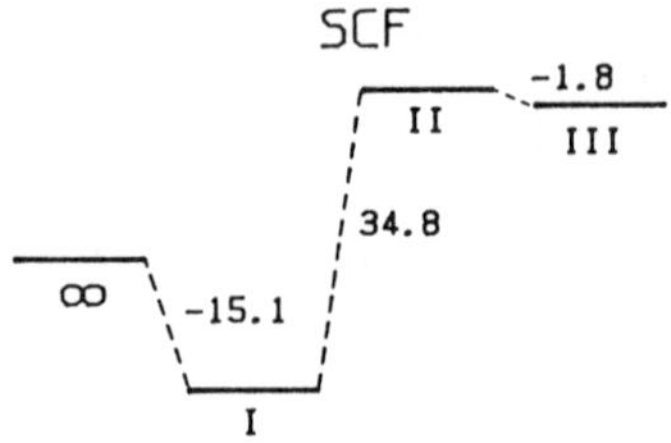

Fig. 4. Relative energies of the model with R = CH_3 for for I, II and III at the SCF level at the optimized proton transfer geometry (4-31G basis set).

With a higher level of accuracy and a more complete geometry relaxation, *in vacuo*, II appears to be merely a point along the straight uphill pathway between I and III.

The results of the STO-3G basis set calculations on the model with R = CH_3 at the 4-31G geometry used in the earlier calculations are reported in Fig. 5. In any case we got either a large barrier or a large energy difference between I and III.

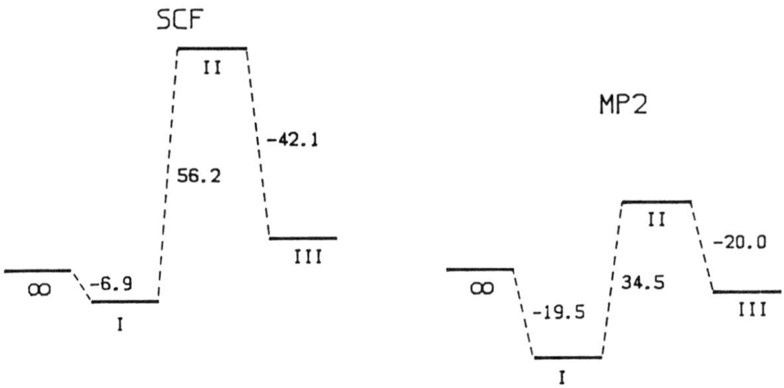

Fig. 5. Relative energies of the model with R=CH_3 at the SCF and MP2 levels for I, II and III, using the STO-3G basis set at the 4-31G geometry.

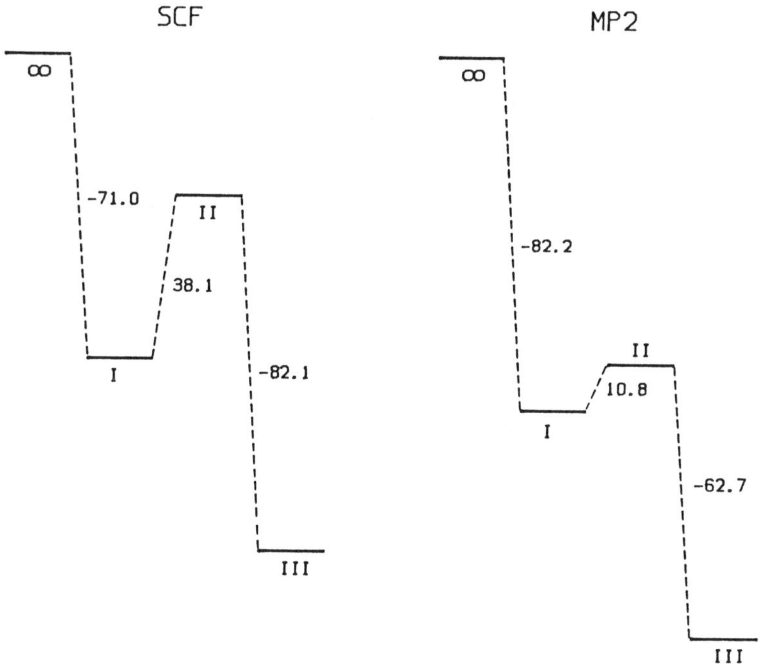

Fig. 6. Relative energies of the model with $R = CH_3$ at the SCF and MP2 levels for I, II and III in the presence of NH_4^+ placed along one of the O lone pairs (R(O...H) = 1.8 Å, <COH = 120°) using the STO-3G basis set at the 4-31G geometry.

To get reasonable results for the first part of the mechanism, it was compulsory to include the environment into the calculations, but first we made a very simple trial to control the feasibility of our study, adding to the quantum mechanical calculations an NH_4^+ group, to simulate the presence of Lys 13. The relative energies at the SCF and MP2 levels are shown in Fig. 6. By comparing Fig. 6 with Fig. 5 one realizes the dramatic effect that charged groups can have in stabilizing III with respect to I. A more realistic description of the key active site residues of TIM was to be derived from the available structures of TIM.

MODEL-BUILT STRUCTURES AND MOLECULAR MECHANICAL SIMULATIONS

X-ray structures of TIM were available both for chicken liver and yeast TIM (YTIM). We turned to these last coordinates as soon as we received them from Alber and Petsko (7), and used the qualitative features of the TIM-DHAP complex model-built at the PS2 to sort those residues (32 in total), that had at least one atom within 8 Å from any atom of DHAP, to be included in the molecular mechanical refinement (8).

The molecular mechanical refinement makes use of an analytical potential of the form:

$$E_{tot} = \sum_{bonds} k_r (r - r_{eq})^2 + \sum_{angles} k_\theta (\theta - \theta_{eq})^2 + \sum_{dihed} (V_n/2)\left[1 + \cos(n\Phi - \gamma)\right]$$

$$+ \sum_{i<j} \left[\frac{A_{ij}}{R_{ij}^{12}} - \frac{B_{ij}}{R_{ij}^{6}} + \frac{q_i q_j}{\varepsilon R_{ij}}\right] + \sum_{H\text{-bonds}} \left[\frac{C_{ij}}{R_{ij}^{12}} - \frac{D_{ij}}{R_{ij}^{10}}\right]$$

to compute the total energy of the system and minimize this functional varying all the parameters in order to make them as close as possible to the reference parameters (9) compatibly with the imposed constraints. One can choose to restrain namely all the C_α carbons of the residues considered in order to keep the backbone in the experimental arrangement or only some of them. We followed the procedure already tested (10) of restraining all C_α carbons at the termini of the chains with a 10 kcal/(mol×Å^2) penalty function to prevent excessive movement of the protein. After a very short minimization to eliminate bad contacts we have model-built three different substrate bound structures (S1, S2 and S3) each one of them with a plausible arrangement of the residues in the active site around DHAP. The last one (S3), that refines to an energy lower than that of the other two structures, was model-built according to the geometry proposed by Alber and Petsko (7) after a difference density analysis of the yeast TIM/DHAP complex and the native enzyme. In this arrangement His 95 and Lys 13 were placed, respectively,

Table I. Results of molecular mechanical refinements.

System	Total energy[a]	Stabilization[a] III vs. I
YTIM/ DHAP (I) (S3)[b]	-465.4	22.9
$YTIMH^+$/ III	-488.3	
DHAP (I)	-16.7	-34.8
III	18.1	
YTIM (mut)[c]/ DHAP (I)	-470.8	72.5
$YTIMH^+$(mut)[d]/ III	-543.3	

a) in kcal/mol; b) the different model-built structures S1 and S2 refined to -443.6 and -464.1 kcal/mol, respectively; c) YTIM with His 95 changed to Gln 95 and energy refined in the presence of the substrate; d) protonated YTIM with His 95 changed to Gln 95 and energy refined in the presence of the enediolate (III).

above and below the plane containing the O=C-C-O-H open ring and, depending on the case, were able to move freely from one O to the other, consistent with the "reversible" character of the isomerization, with Glu 97 also influencing their movement. Given the reasonableness of this structure, we used it to carry out a number of other refinements on the separated reactants and products, as well as on a mutated complex with His 95 substituted by Gln, that we will discuss later on. In Table I we report the energies of a few selected structures.

Interestingly, the molecular mechanical refinements of the isolated species, I and III, give results close to the ab initio ones, but this may be a fortuitous finding. The complex between the protonated YTIM ($YTIMH^+$) and the enediolate (III) is much more favorable (by 23 kcal/mol) than that between YTIM and DHAP (I), confirming (if necessary) the possible effect of polarizing electrophiles on the interaction energy.

ENVIRONMENTAL EFFECT ON THE AB INITIO REACTION MECHANISM

Even though there were no conceptual obstacles in using the charges of all the 32 residues included in the molecular mechanical simulation to describe the field of the enzyme on the substrate, we preferred to use just those expected to most influence the isomerase reaction, i.e., His 95, Lys 13 and Glu 97, separately and jointly, to evaluate the effect produced by each of them. We took into account also the effect of the neutralized phosphate, schematized as a -2/+2 dipole placed along the PO_3^{--} axis. The energies of some of the systems taken into account are reported in Table II.

The role of the enzyme is to stabilize III with respect to I. In fact, considering the H,L,G model as the most appropriate one to represent the environmental effect, we obtain a stabilization by 28.4 kcal/mol at the STO-3G level, and by 26.5 kcal/mol at the 4-31G one.

The inclusion of the charges of the phosphate ion pair destabilizes III relative to I; this is reasonable because the negative charge comes closer to the phosphate in III. It is, however, encouraging that the molecular mechanical simulation, making use of a nonneutralized dianion, stabilizes III with respect to I.

MUTANT ENZYME

We have also included in the *ab initio* calculations, at the same position as His 95, the charges of Gln, to assess the possible effect on the rate of catalysis of this mutation.

We did this test upon the suggestion of G. Petsko, who was interested to assess the importance of a general acid catalyst in the TIM mechanism. In fact, Gln may act, as His, as proton donor and acceptor, but, due to its much larger pKa cannot act as a general acid. The quantum mechanical results (Table II) depicted the mutated enzyme as

Table II. Results[a] of quantum mechanical calculations without and with the incorporation of environment.

System	SCF	MP2[b]	H,L,G[c]	H,l,G[d] (MP2)	H,L,G[e]	G,L,G[f]	G,L,G,P[g]
	in vacuo		native enzyme			mutant enzyme	
STO-RG level							
I	-6.9	-19.5	-30.0	0	-13.7	-29.2	-12.7
II	49.3	15.1	17.2	25.6	43.0	19.1	44.9
III	7.3	-4.9	-44.2	-13.8	-4.5	-40.4	-0.6
III vs. I[h]	-14.2	-14.6	14.2	13.8	-9.2	11.2	-12.1
4-31G level							
I	-14.0(-13.8)	(-13.9)	-45.9	0	-34.3	-44.2	-32.3
II	35.7(32.7)	(18.6)	-3.6	28.3	15.5	-0.9	18.4
III	17.9(11.9)	(8.8)	-40.5	2.4	-7.5	-35.9	-2.7
III vs. I[h]	-31.9(-25.7)	(-22.7)	-5.4	-2.4	-26.8	-8.3	-29.6

a) in kcal/mol: the values in parentheses refer to the model with R=H; b) values including Møller-Plesset (MP2) correlation corrections (6); c) with the inclusion of His 95, Lys 13 and Glu 97 charges; d) the same as c, but including the estimated MP2 correlation corrections; e) including the phosphate as a neutralized dianion (see text); f) with the inclusion of Gln 95 (instead of His 95), Lys 13 and Glu 97 charges; g) the same as e in addition to Gln 95, Lys 13 and Glu 97 charges; h) stabilization energy.

effective as the native one in stabilizing III over I. Also the molecular mechanical results (Table I) were in agreement with the ab initio ones, but when we examined the structures at the Picture System, we noted that Gln had moved during the molecular mechanical minimization to a location different from His and formed an H bond with Glu 165. Gln 95 thus can reduce the catalytic rate with respect to the native TIM by either inhibiting the proton transfer to Glu 165 or by merely not being in position to stabilize the incipient enediolate the way His 95 is. In fact, there is some hint (G. Petsko, private communication) that in the crystal structure of the mutant TIM (His 95 —> Gln 95), Gln 95 moves to form an H bond with Glu 97 rather than with the Glu 165 found in our simulations. This result, if it holds up, would be consistent with our qualitative interpretations based on our simulations, even though the details are not identical (H bond with Glu 97 rather than Glu 165 suggested on the basis of the calculations).

Due to the simplicity of the potential function and the lack of inclusion of solvent, these results must be confirmed by experimental studies, but they are very interesting, having suggested a different result than expected and could be very useful in analyzing possible mutants on proteins whose crystal structure is known, in order either to design new enzymes or to understand enzyme catalysis.

SIMPLE AB INITIO MODEL OF THE ENVIRONMENTAL EFFECT

Exploiting the fact that the enzyme stabilizes III with respect to I, we have designed a very simple model of the enzyme catalysis, putting a point charge in a position such as to equalize the energy of I and III. In the presence of this charge we have then recomputed the whole potential energy hypersurface (reported in Fig. 7), obtaining this time a barrier (Fig. 8) that, when combined with simple Arrhenius theory, gives a rate in very good agreement with the observed catalytic rate. This result is, of

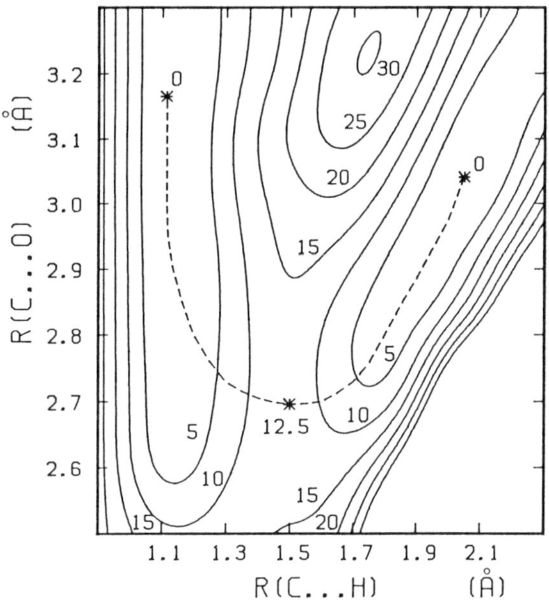

Fig. 7. Potential energy hypersurface in the presence of the simple model of the enzyme (see text) for the conversion I to III as a function of the distance of the proton from the carbon atom and of the total distance between the C and O atoms, at the 4-31G level. Same remarks as in Fig. 3.

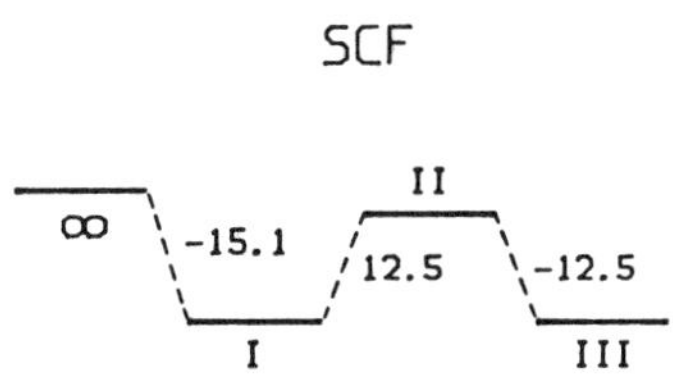

Fig. 8. Relative energies of the model with R = CH_3 at the SCF level for I, II and III in the presence of the simple model of the enzyme, with the 4-31G basis set.

course, somewhat fortuitous but further supports the idea that this is the rate limiting step studied by Straus et al. (11). Substituting Glu 165 with Asp 165 they obtained a reduction in the catalytic rate by about 10^3 such that it becomes the rate limiting step in the catalysis.

We thus used our hypersurface to figure out the effect of removing a single $-CH_2-$ group. The energies of the reactant and the products are rather insensitive to changes in $R(C_1...O)$, so the change of Glu to Asp will have little effect on them. The transition state for proton transfer, on the contrary, is very sensitive to $R(C_1...O)$ and a shortening by 0.3 Å is sufficient to account for the 4 kcal/mol corresponding to the observed decrease in the catalytic rate. We cannot say whether the protein allows the Asp mutant to come as close to DHAP as Glu by straining itself by about 4 kcal/mol or keeps the Asp at $R(C_1...O)$ = 3 Å without strain. In any case, both of these reasons are consistent with an apparently "innocuous" change having a dramatic effect on the rate of enzyme catalysis.

CONCLUSIONS

In summary, we have demonstrated how one can use a combination of computer graphics, molecular mechanics and quantum mechanics to analyze enzyme catalysis. The role of computer graphics is to model build and visualize possible mechanisms for enzyme catalysis and stabilization. The numerical methods, quantum mechanics and molecular mechanics allow a more quantitative analysis of proposed mechanisms. Of course, such methods cannot prove mechanisms, only suggest whether one is plausible or not. Nonetheless, we hope that such approaches will be of use in design of artificial enzymes or in site-specific mutagenesis.

ACKNOWLEDGEMENTS

The work on computer graphics was carried out at the

UCSF Computer Graphics Lab, supported by NIH-RR-1081. P.A.K. is grateful to the NIH for grant GM-29072. This work has been partly supported by the Italian C.N.R. The plots have been drawn at ICQEM with the HIPLOT DMP-29 pen plotter (P. Palla, system manager).

APPENDIX

In this study we used the following programs: GAUSSIAN 80 UCSF (12), running on the UCSF Structural Biology VMS VAX 11/780 and GAUSSIAN 80 (13), running on the IBM 3033 at the CNUCE Computing Center (Pisa) for the quantum mechanical calculations; AMBER (8), running on the VAX 11/780 for the molecular mechanical calculations, and CHEM (14), running on the UCSF Computer Graphics Lab VAX 11/750 and Evans & Sutherland Picture System 2 for the computer model building.

REFERENCES

1. Alagona, G., Desmeules, P., Ghio, C. and Kollman, P.A. (1984) J. Amer. Chem. Soc. 106, 3623-3632.
2. Alagona, G., Ghio, C. and Kollman, P.A. (1986) J. Mol. Biol. 191, 23-27.
3. A review of the various quantum mechanical methods currently employed to incorporate the environment into the calculations appeared recently: Drummond, M.L.J. (1986) Prog. Biophys. Molec. Biol. 47, 1-29.
4. Warshel, A. and Sussman, F. (1986) Proc. Natl. Acad. Sci. U.S.A. 83, 3806-3810.
5. Weiner, S.J., Seibel, B.L. and Kollman, P.A. (1986) Proc. Natl. Acad. Sci. U.S.A. 83, 649-653.
6. Binkley, J.S. and Pople, J.A. (1975) Int. J. Quantum Chem. 9, 229-236.
7. Alber, T. and Petsko, G.A. J. Mol. Biol., submitted for publication; Alber, T. and Petsko, G.A. Biochemistry, submitted for publication.
8. Weiner, P. and Kollman, P.A., (1981) J. Comput. Chem. 2, 287-303.

9. Weiner, S.J., Kollman, P.A., Case, D.A., Singh, U.C., Ghio, C., Alagona, G., Profeta, S., Jr. and Weiner, P. (1984) J. Amer. Chem. Soc. 106, 765-784.
10. Wipff, G., Dearing, A., Weiner, P., Blaney, J. and Kollman, P. (1983) J. Amer. Chem. Soc. 105, 997-1005.
11. Straus, D., Raines, R., Kawashima, E., Knowles, J.R. and Gilbert, W. (1985) Proc. Natl. Acad. Sci. U.S.A. 82, 2272-2276.
12. Singh, U.C. and Kollman, P.A. (1982) QCPE Bull. No. 2, Program No. 446.
13. Whiteside, R., Krishman, R., Seegar, R., De Frees, D., Schlegel, H., Binkley, J., Topiol, S., Kahn, J. and Pople, J; (1980), Gaussian 80, QCPE Program No. 406.
14. CHEM, a program for molecular manipulation on the Evans & Sutherland PS2, written by A. Dearing at UCSF.

SYNTHETIC PEPTIDES AND THE DESIGN OF PEPTIDE AND PROTEIN RECOGNITION SURFACES

Irwin M. Chaiken, Shoji Ando, Yechiel Shai, Giorgio Fassina, and Xun Liang*
National Institute of Diabetes and Digestive and Kidney Diseases, National Institutes of Health, Maryland, USA, and
*Department of Chemistry, Nankai University, Tianjin, Peoples Republic of China

I. INTRODUCTION

Macromolecular recognition is a unifying theme in biology and biotechnology and understanding the mechanisms underlying this phenomenon remain important goals of biochemical study. The structural nature of recognition surfaces for an increasing array of macromolecules, especially proteins, can now be described at high resolution thanks largely to diffraction analysis of their crystals. Such protein structures offer an improved opportunity to focus solution biochemical experiments in order to define the forces responsible for their surface interactions and consequent functions. Moreover, improved understanding of protein as well as peptide interactions brings with it the ability increasingly to use these phenomena in technology, for example, as vehicles for separation, enzyme regulation, immunodiagnostics and therapy, the redesign of native macromolecules to produce new functions and the *de novo* design of non-biological surrogates. Given an awareness of the importance of peptide and protein recognition and the need to improve our understanding of its common mechanistic principles, finding experimental approaches to study recognition surfaces remains a pre-occupation of peptide and protein chemistry.

One approach with a rich history and yet important continuing usefulness to study recognition mechanisms is synthetic mutation (Fig. 1). The sequence engineering of peptides and proteins that was first accomplished by total peptide synthesis and semisynthesis has now been complemented by the powerful technology of recombinant DNA mutagenesis. Direct peptide synthesis offers a flexible tool to examine the mechanistic roles of amino acid residues in peptide and protein recognition surfaces and to test theoretical concepts of what structural elements actually are most needed to promote and control recognition. The major realm of chemical synthesis has been that of relatively small biologically active polypeptides and their mutation not only with common amino acid residues but also with non-DNA encoded residues such as N-methylated and D-amino acids. While total chemical synthesis has suffered from limits on the size of polypeptides that can be made routinely (a feature necessary for elaborating a significant series of sequence mutants), semisynthesis overcomes such size limitations, by permitting relatively small, synthesized peptides to be recombined with fragments of native polypeptides and proteins

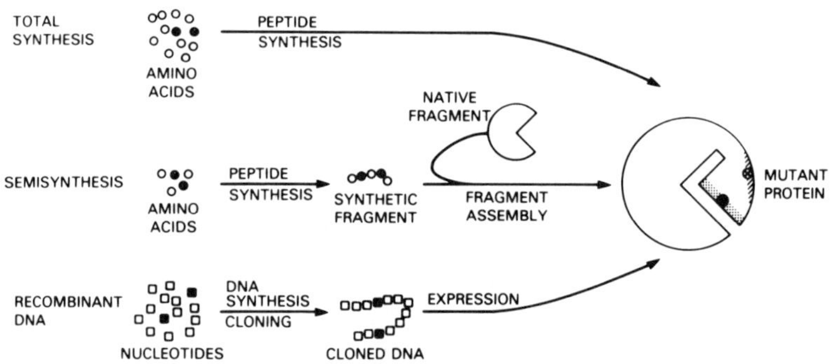

Fig. 1. Schematic diagram of synthetic design of protein mutants with sequence-engineered recognition surfaces. Intramolecular (dotted) and intermolecular (cross-hatched) surfaces are depicted. Sites of amino acid replacement are designated by ● and ⊗, while nucleotide base sequence replacements that give rise to the amino acid replacements are designated by ■ and ⊠, respectively.

to produce functional hybrids. Thus, synthesis of mutants can be achieved for the sequence engineering of increasingly larger proteins. The most profound liberation from size constraints has come, however, not from semisynthesis but from recombinant DNA approaches. The latter enable sequence mutation and design for a rapidly expanding range of large polypeptides and proteins by mutation of cloned genes.

In truth, however, whatever the particular experimental method used for sequence engineering, the conceptual questions asked are generally the same. Current experimental results suggest a growing usefulness of all synthetic approaches in the overall attempt to unravel the rules of protein folding and recognition. It is the purpose of this paper to review some of the currently held views of peptide and protein recognition as they have been addressed by peptide synthesis and semisynthesis. The emphasis obviously is personalized, reflecting as it does the experience of the authors. For interested readers, recent discussions of studies of macromolecular recognition mechanisms using recombinant DNA and natural genetic sequence variation (1,2) as well as earlier discussions for peptide synthesis and semisynthesis (3,4) provide complementary insights into the overall trends in using sequence variation to characterize and re-design peptide and protein recognition surfaces.

II. SEQUENCE SIMPLIFICATION AND CHARACTERIZATION OF THE S-PEPTIDE/S-PROTEIN RECOGNITION PROCESS

We have used peptide synthesis to examine underlying principles which govern the peptide/protein interaction in the case of bovine pancreatic ribonuclease-S (RNase-S). Sequence design experiments with this "prototype" system benefit from a large body of data from evolutionary sequence variation, crystallography, and, for the S-Peptide component, one-site-at-a-time synthetic mutation (3,5-7). Based on these previously-obtained data, especially those relating to stabilization of S-Peptide/S-Protein contact, we asked (8) whether one could "decode" this recognition process by hypothesizing how sequence information is used to produce contact. We proposed (Fig. 2) that, in S-Peptide, the most crucial elements of structure for functional recognition are an α-helical conformational framework provided by residues 3-13, and 2 contact elements, the side chains of Phe8 and Met13, positioned by the α-helical framework to match a complementary recognition surface in folded S-Protein. We reasoned further that the α-helix, when oriented by Phe8 and Met13 contact on the S-Protein surface, was sufficient to place the essential catalytic residue His12 in the RNase-S catalytic site.

A

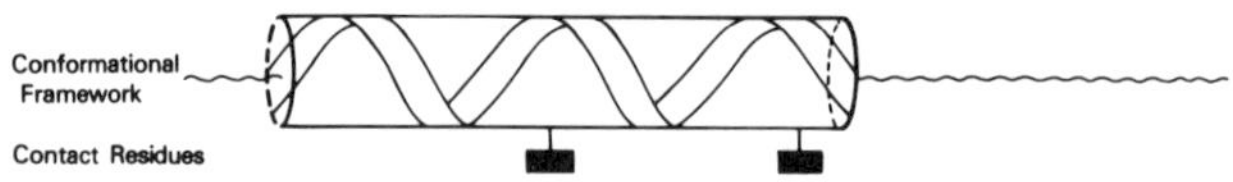

B

	Sequence	Activity at Saturation (%)	K_d (at [S] = o) (M)
Parent Model	H_2N- Ala Glu Ala Ala Ala Ala Lys Phe Ala Arg Ala His Met Ala Ala- COOH	45	3.3×10^{-6} $5 \times 10^{-6*}$
Ala 8 Mutant	——————————Ala——————————	∼5	$\sim 1 \times 10^{-3*}$
Ala 13 Mutant	————————————————Ala————	∼5	$> 10^{-3*}$
Simplified Model	H_2N- Glu Ala Ala Ala Ala Ala Ala Phe Ala Ala Ala His Met Glu Ala Ala Ala Ala Ala Ala- COOH	15	$\sim 5 \times 10^{-4*}$
Native	H_2N- Lys Glu Thr Ala Ala Ala Lys Phe Glu Arg Gln His Met Asp Ser Ser Thr Ser Ala Ala- COOH	(100)	6.3×10^{-7}

Fig. 2. Overall logic and redesign by sequence simplification of ribonuclease S-Peptide recognition surface for S-Protein (8-10). A: Diagram depicting critical elements of structure for S-Peptide recognition. These include an α-helical conformational framework of residues 3-13 and two contact side chains at Phe8 and Met13. B: Sequence of modeled S-Peptides with α-helical frames formed largely with a poly-Ala core and the contact residues Phe8 and Met13 either retained or placed by Ala residues. Activities at saturation are the maximum enzymatic activities, relative to that of native S-Peptide/S-Protein complex, produced upon titrating native S-Protein with sequence modeled peptide with 2',3'-cyclic cytidine monophosphate as substrate at 0.1M Tris HCl, pH 7.13 (8,9). Values of K_d are for dissociation constants of native or model S-Peptide binding to S-Protein as determined by analytical affinity chromatography in the absence of substrate, by analyzing the effect of S-Peptide on elution of S-Protein on immobilized S-Peptide or, where denoted by "*", peptide elution on immobilized S-Protein (11).

On this basis, we have prepared (8-10) a series of synthetic, sequence-modeled S-Peptides in which much of the sequence was replaced by multiple Ala residues to provide a sequence simplified poly-Ala α-helical frame; Phe8 and Met13 were preserved to "recognize" the S-Protein surface and His12 also was retained in order to use the appearance of enzymatic activity as at least one diagnostic of functional peptide/protein interaction. In the parent model S-Peptide, Glu2, Lys7, and Arg10 from the native sequence also were retained, the first and third to provide a possible helix-stabilizing salt bridge (suggested from earlier crystallographic data) and the second residue since some experiments had suggested its possible importance in RNase function. However, in more simplified models these three residues were eliminated without total loss of S-Protein recognition, although the affinities of these latter models were certainly weaker than that of the parent model.

The structural properties of the parent model semisynthetic ribonuclease-S species formed by sequence-simplified S-Peptide were characterized both in solution and in crystals (8-10,12). The increase in enzyme activity which occurred when parent model was added to S-Protein showed that this peptide could mimic native S-Peptide, with the peptide/protein affinity in the presence of substrate being about one order of magnitude weaker, and the activity at peptide saturation being about 45%, versus native complex (Fig. 2). Moreover, by quantitative affinity chromatographic analysis, measuring the binding of model S-Peptide to S-Protein by its effect on the elution of S-Protein on immobilized S-Peptide, model S-Peptide recognition of S-Protein in the absence of substrate also was found to be related to that with native S-Peptide by close to a one order of magnitude decrease (Fig. 2).

While the above data confirm that the model peptide has enough sequence information to mimic native S-Peptide effectively, the most graphic verification of structural mimickry upon sequence simplification was obtained (12) by X-ray diffraction analysis of the model semisynthetic complex. As shown in Fig. 3, model and native S-Peptides assume virtually the same helical structure and orientation with respect to S-Protein, represented as the dot surface, in native-sequence and model semisynthetic RNase-S, respectively. And, the Phe8 and Met13 side chains make contact with the same hydrophobic area (brighter dots below side chains) on the S-Protein surface. It is this orientation which places His12 in the RNase catalytic site, in the back of the S-Peptide in the view of Fig. 2 and close to His119 (not shown) from S-Protein.

The results of semisynthetic RNase-S sequence modeling validate the idea of characterizing recognition processes of peptides and proteins by a sequence simplification approach that retains the core elements of conformational framework and surface recognition contacts. The model complex also provides an opportunity to use synthetic redesign to modify the recognition process itself, by changing elements of surface contact. We have tested this idea (10) by changing the sequence of parent model S-Peptide either to eliminate Phe8 or Met13 (replacing by Ala in each case) or to alter the stability of the Phe8

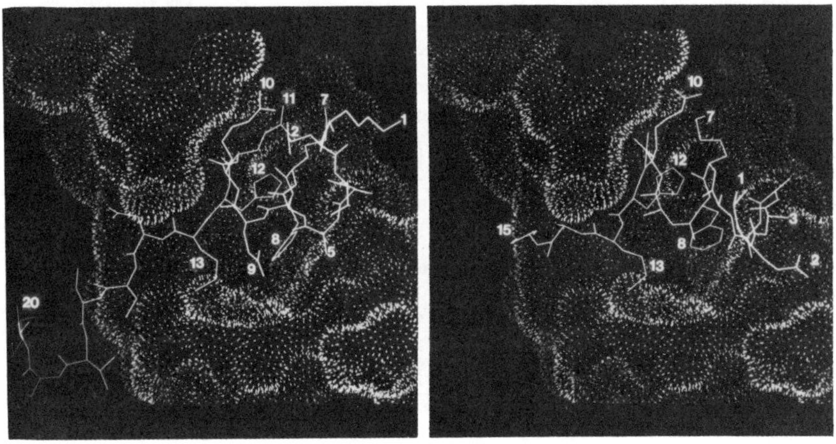

Fig. 3. Comparison of structures of native S-Peptide (left) and parent model S-Peptide (right) in complexes with S-Protein, as determined by X-ray diffraction analysis (12). Peptides are shown as stick diagrams with several residues numbered for reference. The S-Protein is denoted as a dot surface. (Adapted from Ref. 12 with permission).

contact by replacing Ala9 by Pro. All of these replacements destabilize S-Peptide/S-Protein recognition, as expected, verifying the overall view that positions 8 and 13 are the primary contact sites. These results should make it possible to put new residues in positions 8 and 13, and perhaps elsewhere, both to "reinstate" high affinity binding to S-Protein and perhaps even to willfully alter the sensitivity of binding towards such solution conditions as pH and temperature.

III. STUDYING RECOGNITION MECHANISMS IN BIOSYNTHETIC PATHWAYS FOR NEUROENDOCRINE PEPTIDES

With the RNase S-Peptide data as a stimulus, the sequence simplification approach to synthetic design of peptides and proteins can be seen as a general approach to evaluate the nature and role of molecular recognition events in biological systems. We have begun to use synthetic redesign to study multimolecular assembly mechanisms which occur in the neuroendocrine biosynthetic pathways of oxytocin and vasopressin. These peptide hormones, as generally for neuropeptides such as adrenocorticotropic hormone, endorphin, and enkephalin, to name only a few, are produced in cellular pathways (12,13) as depicted in Fig. 4. Several molecular conversion/translocation events occur in such pathways, including synthesis of protein precursors, often containing sequence domains for multiple active polypeptides; packaging at high concentrations into secretory granules; enzymatic processing of precursors to mature peptides during granule transport and storage; and eventual exocytotic co-release of active peptide contents from granules for transport to post-synaptic or peripheral sites. Little is known of the molecular organization of precursors and products in the concentrated milieu of the secretory granules (15) and the impact of this organization on the molecular transformations of precursor proteins to active peptides. Yet, from an awareness of the properties of polypeptides and proteins generally, it is reasonable to expect that precursors would assemble into folded, and perhaps non-covalently interacting, forms and that these molecular assembly events could have an impact on the control of peptide production. We were interested to investigate the nature and role of assembly of neuroendocrine precursor proteins in peptide-producing pathways.

In the case of the neuroendocrine peptides oxytocin and vasopressin, there are several clues that the biosynthetic precursors assemble into folded, well-ordered forms. The precursor of each of these peptides (Fig. 5) contains a hormone sequence domain (9 residues, 1 intradomain disulfide) at the amino terminus, a Gly-Lys-Arg tripeptide spacer/enzymatic processing re-

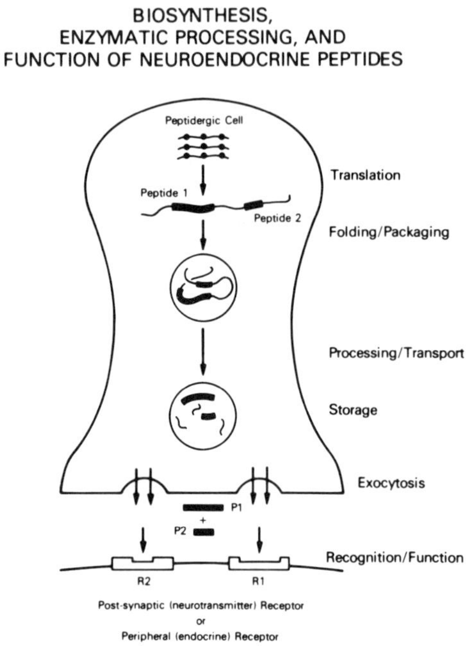

Fig. 4. Schematic diagram of neuroendocrine peptide/protein pathways. Depicted are the biosynthesis, folding, and granule packaging of precursors containing multiple activity domains; enzymatic processing of precursors to produce mature, active polypeptides; axonal transport and storage of mature granules; exocytotic release of neuroendocrine polypeptides; and ultimate action of active peptides at postsynaptic or peripheral target receptors (Fig. taken from Ref. 14 with permission).

gion, a sequence domain for a coordinately synthesized neurophysin (small protein of 10,000 daltons, containing 7 intra-domain disulfides), and a variable carboxyl terminal extension, either His in the oxytocin precursor or Arg-glycopeptide in the vasopressin precursor (Fig. 5). Upon completion of enzymatic processing, the mature sequence domains (NP and H) interact to form noncovalent complexes; and, these 1:1 complexes can self-associate (19-23). Thus, it may be hypothesized (Fig. 5) that the precursors may also fold, with hormone and neurophysin domains interacting intramolecularly, and that these folded precursors may self-associate into at least dimers (19,24,25). Spontaneous precursor folding would also explain the genesis of native disulfides, which cannot form spontaneously in mature neurophysin itself (26,27).

To test the hypothesis of neuroendocrine precursor assembly, we chose the synthetic redesign approach using

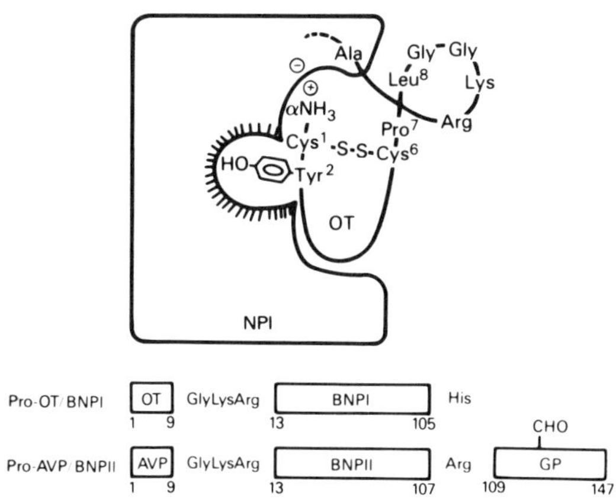

Fig. 5. Top: Schematic view of the intramolecular interaction between hormone and NP domains in OT/BNPI biosynthetic precursor, emphasizing the role of the α-amino and Tyr 2 side chain of the hormone in this interaction and the sequence position of the inter-domain sequence. Bottom: The primary structures of pro-OT/BNPI and pro-AVP/BNPII from cDNA sequence analysis (16,17) (Adapted from Ref.18).

the simpler oxytocin precursor as the first case to study. We made precursor and sequence mutants by the semisynthetic route (18,25), synthesizing the oxytocinyl-Gly-Lys-Arg segment containing the hormone domain and then, after suitable side chain protection, chemically coupling this piece to the isolated (ε-amino group blocked) bovine pituitary neurophysin I (BNPI). In addition to the normal sequence semisynthetic precursor, pro^{S}-OT/BNPI, two mutants were made by synthesis and coupling of oxytocinyl-Gly-Lys-Arg mutants to BNPI (18). In one mutant, the $N^{\alpha 1}$-acetyl,Ala2 mutant or pro^{S}-AA/BNPI, two structural elements were eliminated which were predicted to be essential for domain-domain contact, namely the positive charge at the α-amino group and the Tyr2 side chain. This mutant was used to test the dependence of precursor self-association on domain/domain interaction. In the second mutant, the D-Pro7, D-Leu8 or pro^{S}-DD/BNPI species, the role of interdomain linker conformation in intramolecular folding was tested by changing the geometry

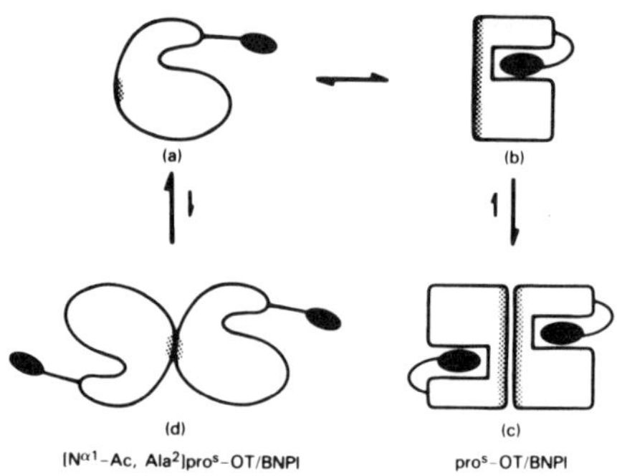

Fig. 6. Mechanism of self-assembly tested with semisynthetic pro-OT/BNPI and precursor mutants, emphasizing the expected linkage between intramolecular domain recognition and precursor self-association. The ring portion of hormone sequence is shown as a blackened oval, which is connected to NP domain by a peptide spacer. In this view, upon hormone domain binding to the complementary surface of the NP domain, the conformational form of the NP domain is stabilized in which the self-association surface (dotted area) is in the high affinity state (Adapted from Ref. 18).

in the sequence region between the two internal recognition domains of hormone disulfide loop and the complementary BNPI surface.

Experimentally, two signals were followed to measure the assembly properties of pro^{S}-OT/BNPI and its mutants. The first was affinity chromatographic behavior on immobilized neurophysin and hormone analogue affinity matrices (28,29). It was expected that, if precursors folded with domains interacting intramolecularly and exhibited enhanced self-association potential as a result (Fig. 6), the precursors would not be retarded on the peptide matrix but would have a high affinity for neurophysin matrix similar to that of liganded neurophysin. In contrast, if intramolecular domain folding did not occur, the precursor would be expected to bind to the peptide matrix, via the unoccupied neurophysin domain, with an affinity similar to that of neurophysin itself but to bind to neurophysin matrix with the low affinity expected of non-liganded neurophysin. These above expectations were borne out as shown by the representative data of Fig. 7 A-D. The intramolecular domain-domain interaction of the semisynthetic precursors also was measured by circular dichroism, using the increased negative ellipticity at 291 nm versus that at 280 nm as a diagnostic. This differential ellipticity increase was shown previously (30) to occur upon hormone-neurophysin noncovalent interaction. The CD results (Fig. 7 E and F) confirm the interaction properties schematically depicted in Fig. 6. Here, pro^{S}-OT/BNPI shows the diagnostic shoulder at 291 nm, while pro^{S}-AA/BNPI does not. Interestingly, pro^{S}-DD/BNPI also has a θ_{291nm} indicative of native-like assembly up to 40°C; but, elevating temperature above 40° leads to a progressive disappearance of the signal even though that for pro^{S}-OT/BNPI is undiminished at least to 60°C. Thus, the conformation of the sequence region between hormone disulfide domain and neurophysin domain is contributory, though relatively weakly, to the interaction of domain surfaces within the precursor.

The precursor assembly properties as revealed by semisynthetic sequence design, taken together with a rather large body of data on noncovalent interactions of hormones and neurophysins (19-23,29), lead to the view of molecular interactions in the hormone/neurophysin pathway

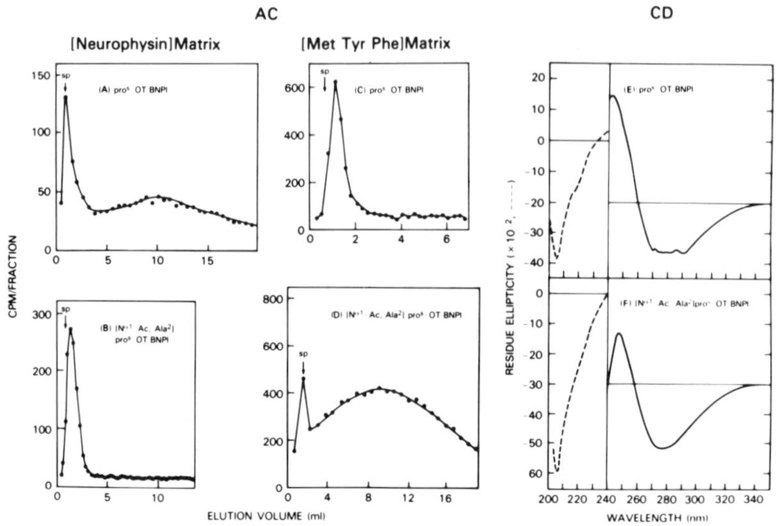

Fig. 7. Analyses of intramolecular and intermolecular assembly properties of semisynthetic pro^S-OT/BNPI and the semisynthetic mutant [$N^{\alpha1}$-Ac, Ala^2]pro^S-OT/BNPI. A and B: Analytical affinity chromatographic elutions of semisynthetic precursor and mutant on [Met-Tyr-Phe]Affi-Gel 102. Zones containing radiolabeled semisynthetic precursors were fractionated on a [Met-Tyr-Phe]Affi-Gel 102 column of 8.5 µmol peptide/ml of bed volume; 353 µl bed volume; V_o = 750 µl; V_m = 420 µl; elution with 0.4 M ammonium acetate buffer pH 5.7 containing 0.5% BSA. SP indicates the elution position of the non-interacting ribonuclease S-protein (monitored by absorbance at 280 nm) and is taken as V_o. C and D: Analytical affinity chromatography of semisynthetic precursors on a [BNPII]Sepharose column of 52 nmole of BNPII/ml of bed volume; 353 µl bed volume; V_o = 692 µl; V_m = 445 µl; elution with 0.4 M ammonium acetate, pH 5.7, containing 0.5% BSA. The peak at V_o (SP) is a small amount of inactive radiolabeled polypeptide. E and F: Circular dichroism spectra of semisynthetic precursors. Conditions: pro^S-OT/BNPI (7 µM) and [$N^{\alpha1}$-Ac, Ala^2]pro^SOT/BNPI (20 µM), each in 10 mM sodium phosphate, pH 5.7. Data in the far ultraviolet region (200-240 nm) were calculated as ellipticity per amino acid residue. Data in near ultraviolet region (240-350 nm) were calculated as ellipticity per mole of protein (all data taken from Ref. 18).

shown in Fig. 8. Here, both precursors and processing products are viewed as assembling into liganded dimers. This potential also has been confirmed with pro^{S}-AVP/BNPII (31). Given the high concentrations (at least 0.1 mM) of precursors and their processing products in peptide secretory granules (15, Fig. 4), it is likely that the molecular forms which exist in granules are the self-associated species and that the substrates of enzymatic processing are assembled, folded precursor aggregates. This further suggests that enzymatic processing may well be regulated by ordered precursor assembly, a phenomenon which is under current study in our laboratory (32).

Finally, the semisynthetic approach offers a potentially convenient tool to use recognition surface redesign to learn how hormone-neurophysin and neurophysin-neurophysin interactions are linked structurally. The

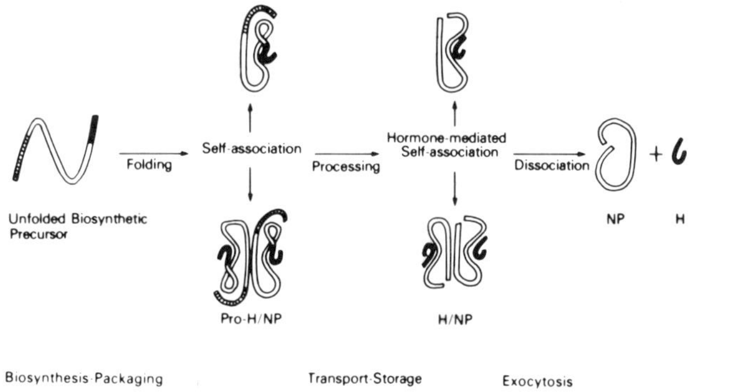

Fig. 8. Schematic model depicting relationship of biosynthetic precursor structure to molecular events occurring in neurohypophysial hormone/neurophysin biosynthesis. The filled and open lines denote hormone and neurophysin sequence domains, respectively. The cross-hatched line represents the C-terminal glycopeptide occurring in the Arg8 vasopressin/neurophysin precursor. Folding of the precursor leads to high affinity of self-association through the NP domains of the precursors. The NP-NP and H-NP interaction surfaces are retained after enzymatic processing, which leads to formation of noncovalent complexes between H and NP in secretory granules until released exocytotically. (Fig. taken from Ref. 14 with permission.

cooperativity between these interactions is apparent in both noncovalent complexes and precursors. While the currently described experiments with precursors show that the interaction of hormone disulfide domain with neurophysin domain is the "trigger" for high affinity precursor self-association, the intramolecular structural network which conveys the message between the two sites is yet to be defined. Characterizing the structural basis of cooperativity in the rather simple system of the oxytocin and vasopressin precursors, including current efforts to sequence-simplify the hormone domain and spacer regions using the synthetic approach, may also provide unifying ideas about structural linkage and cooperativity in interacting macromolecular systems generally.

IV. ORTHODOXY IN PEPTIDE AND PROTEIN RECOGNITION: DOES IT ALWAYS FIT?

The recognition mechanisms in RNase-S and pro^{S}-OT/BNPI both reflect a rather widely-accepted orthodox view of molecular recognition. In both cases, compact, folded molecular domains interact by recognition surfaces involving a relatively few structural elements which are positioned at complementary surfaces by the conformational frameworks of the interacting domains. In the S-Peptide case, the conformational frame is an α-helix and the contact residues are Phe8 and Met13 (Figs. 2,3). In the oxytocin precursor case, the conformational frame can be viewed as the 6-residue disulfide loop and the contacts to neurophysin are an α-amino and Tyr2 side chain (Fig. 2,5,6,8). This simple view of intramolecular and intermolecular recognition no doubt fits many macromolecular systems whose structures are known and explains the result of sequence variation and mutation in many of these where such data are available.

Nonetheless, recent experiments with anti-sense peptides demand a more sober consideration of whether the orthodox view of molecular recognition is all-encompassing. Blalock and his colleagues (33-35) have observed that synthetic peptides encoded in the base sequence of anti-sense DNA (Fig. 9) are able to bind to the naturally occurring sense-peptides. The phenomenon, first observed (33) with ACTH (sense peptide) and the anti-sense peptide

"HTCA", could be considered a reflection of complementarity of recogniton surfaces of compactly folded sense and anti-sense peptides, a view which would predict that the anti-sense peptide might mimic the receptor of sense peptide. Indeed, antibodies made against the anti-sense HTCA were found to bind to cell membrane receptors of ACTH (33,34). While sequence comparison of sense and anti-sense peptides has suggested a matchup in the patterns of hydropathic properties of the residues (35), the question nonetheless has remained of whether anti-sense recognition of sense peptides fits the orthodox model of macromolecular recognition, i.e. that the peptide partners fold into compact conformations with well-defined and complementary recognition surfaces (e.g. Figs. 2,5). Nonetheless, the potential to use anti-sense peptides as a guide to design synthetic molecules which recognize natural peptides and proteins has impressive implications

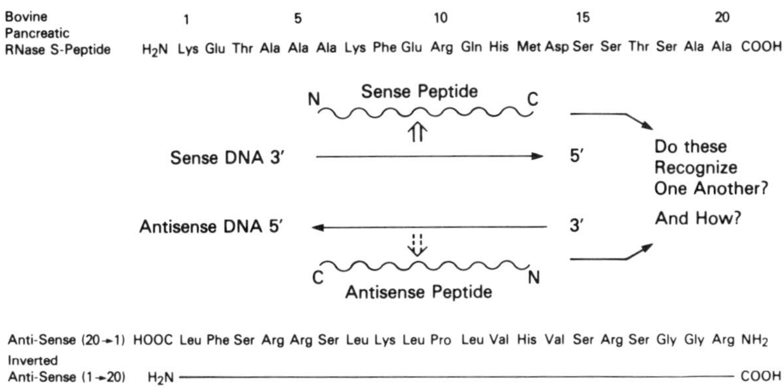

Fig. 9. Scheme depicting relationship of anti-sense peptide to sense peptide and the S-Peptide related experimental system devised to test this relationship. Anti-sense peptide is the sequence encoded in anti-sense (or complementary) DNA, while sense peptide is the native sequence encoded in the sense (coding) strand of DNA. The diagram shows the sequences of ribonuclease S-Peptide and two synthetic anti-sense peptides, one (A.S. (20→1)) corresponding to anti-sense DNA read in a normal 3' to 5' direction and the other (inverted anti-sense (or I.A.S. (1→20)) with the same sequence as A.S. but with amino to carboxyl terminal residue orientation reversed.

both for the study of forces which produce peptide and protein recognition in biology and for biotechnological uses as affinity agents in separation and analysis.

We recently have begun an investigation of the occurrence and nature of sense/anti-sense peptide recognition (11) using RNase S-Peptide as sense peptide, solid phase synthetized antisense peptides, and the experimental method of analytical affinity chromatography (Fig. 10; for general reviews, see Refs. 29,36-38) to measure peptide interactions. The S-Peptide system was chosen since

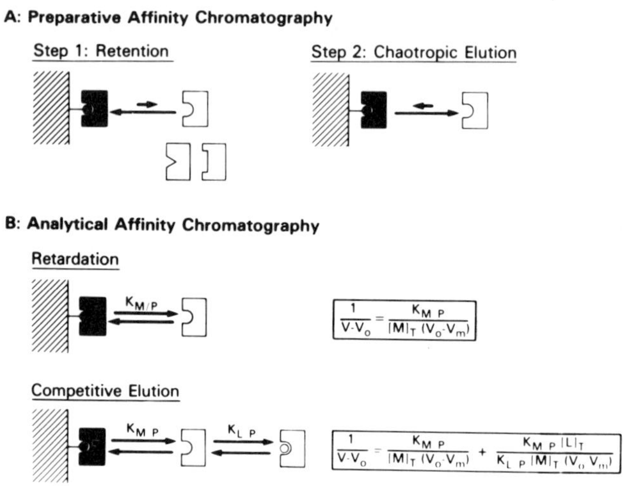

Fig. 10. Diagram comparing preparative (A) and analytical (B) affinity chromatography. The latter method, with isocratic elution either with (B-lower) or without (B-upper) competitor in buffer, allows measurement of values of V (experimental elution volumes) and consequent use of equations shown to calculate $K_{M/P}$ and $K_{L/P}$, the dissociation constants, respectively, of mobile peptide (P)/immobilized ligand (M) complex and of mobile peptide (P)/soluble competitor (L) complex. For the anti-sense/sense peptide experiments, M and L are immobilized and competitor sense S-Peptide, respectively, while P is anti-sense or inverted anti-sense peptide. In the equations relating V to $K_{M/P}$ and $K_{L/P}$ (36), V_o is unretarded elution volume, V_m is the accessible volume (outside pores of affinity support) and $[M]_T$ is the concentration of immobilized ligand. (Fig. adapted from Ref. 37).

there is a rather deep understanding available of native S-Peptide conformational tendencies, alone and when it interacts with S-Protein; thus the extent of contribution of conformation to sense/anti-sense binding could be assessed from the starting point of reasonable background information, at least for the sense peptide component. Furthermore, the analytical affinity chromatographic approach was chosen to measure peptide interactions since it is a method by which molecular interaction can be measured directly, without depending on observing effects of interaction such as an activity or spectroscopic change, neither of which could be predicted to occur with anti-sense peptide binding. In addition, when sense peptide is the immobilized molecule (M), chromatographic elution of anti-sense peptide (P) either alone (Fig. 10B, upper) or in the presence of mobile sense peptide competitor (L, Fig. 10B, lower) allows quantitative values of K_d's for anti-sense/sense interaction to be determined, using isocratic elution, for the immobilized sense/soluble anti-sense interaction ($K_{M/P}$) and for soluble sense/soluble anti-sense interaction ($K_{L/P}$). Finally, immobilized S-Peptide (the sense component) can be made which is functionally active as judged by the K_d of S-Protein binding to it, thus establishing appropriate chromatographic conditions for anti-sense peptide binding studies.

When anti-sense S-peptide, the sequence encoded by putative anti-sense DNA of S-Peptide read 3' to 5' in DNA for amino to carboxyl in peptide, was eluted on immobilized S-Peptide, retardation was observed; and the retardation was established to be peptide-directed by competition with soluble sense S-Peptide (Fig. 11A). The K_d's for binding of anti-sense peptide to immobilized and soluble sense peptide, $K_{M/P}$ and $K_{L/P}$ respectively, were calculated as given in the figure, using the equation shown in Fig. 10B for competitive elution.

The results not only confirm that anti-sense/sense peptide binding can occur but provide a quantitative measure of the extent of interaction. Strikingly, when the experiment was repeated with inverted anti-sense peptide, the same sequence as anti-sense S-Peptide but with a reversed amino-to-carboxyl orientation (amino to carboxyl of peptide corresponding to 5' to 3' for anti-sense DNA), binding to sense peptide was observed here also, and with an affinity only slightly lower than that for normally oriented anti-sense peptide (Fig. 11B). Since it can be hypothesized that anti-sense and inverted anti-sense peptides do not have the same three-dimensional structure, a tentative conclusion from the results of Fig. 11 is that anti-sense and sense peptides bind by

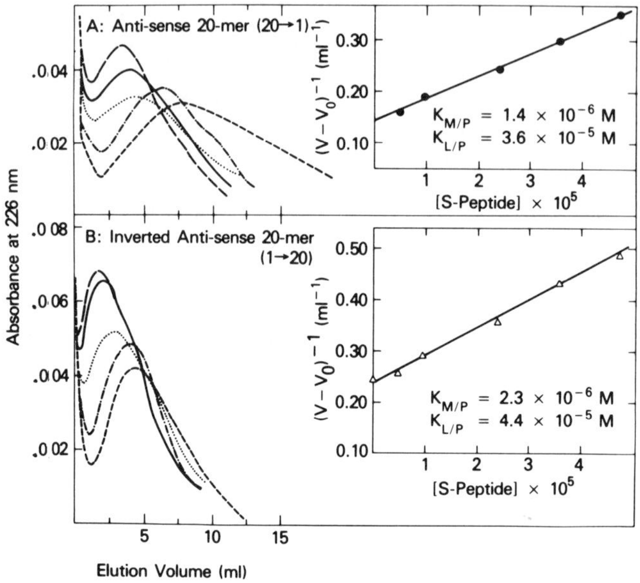

Fig. 11. Competitive zonal elution HPAC analysis of anti-sense 20 mer peptide binding to RNase S-peptide. Zones of 7.6 nmoles of either (A) A.S.(20→1) or (B) I.A.S.-(1→20) in 10 μl were eluted on a 0.32 ml bed volume column of silica-immobilized S-peptide, $[M]_T = 9.1 \times 10^{-5}$M, in 0.2 M NH_4OAc, pH 5.7, containing the following molar concentrations of soluble S-peptide: ---- 0.47×10^{-5}; —·— 0.95×10^{-5}; ...... 2.39×10^{-5}; —— 3.58×10^{-5}; and — — 4.78×10^{-5}. Elution profiles are shown in the main figures; the insets show the variation of V with soluble competitor concentration plotted as $1/(V-V_o)$ vs. S-Peptide according to the equation in Fig. 10B lower. In (B), the elution profile for [S-Peptide] = 0 is omitted to avoid clutter, but the value of $1/(V-V_o)$ is given in the inset along with the other data obtained. The values of $K_{M/P}$ and $K_{L/P}$ determined from the data (using the equation in Fig. 10B lower) are given in the insets of (A) and (B). (Taken from Ref. 11 with permission).

recognition surfaces which do not require a single type of compact, folded structure as visualized in Fig. 12 for RNaseS and pro^S-OT/BNPI. One way recognition could occur here is by a matchup of sequence in elongated peptide forms. This latter view is supported by the finding that progressive shortening of anti-sense peptides leads to a stepwise decrease in sense peptide affinity but not the discontinuous decrease one might expect if only a few residues were responsible for most of the binding energy. Finally, analysis of the stoichiometry

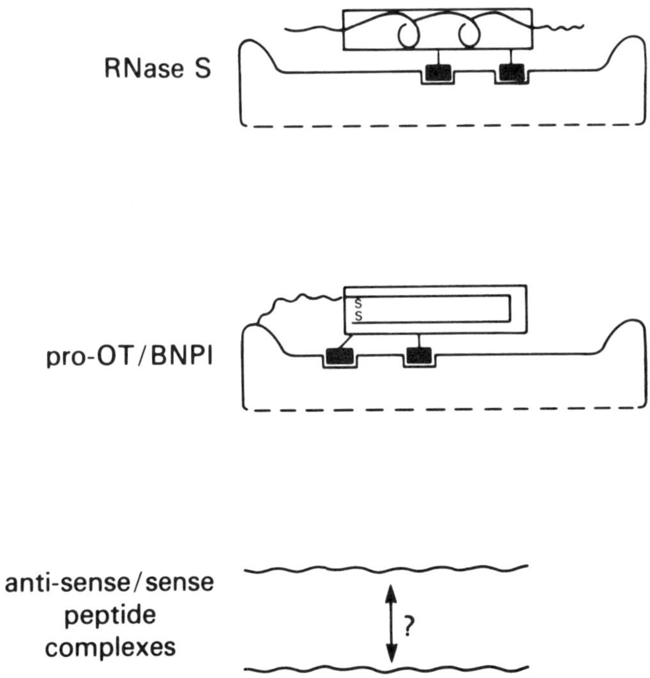

Fig. 12. Scheme comparing the modes of binding of S-Peptide with S-Protein in RNase-S, hormone with neurophysin domains in pro-OT/BNPI, and anti-sense with sense peptide. The first two can be depicted as occurring by interaction of compactly folded molecular species at well-defined recognition surfaces composed of a relatively few contact residues placed sterically by well-defined conformational frameworks (see Fig. 2). The mode of anti-sense/sense peptide binding is not well-understood but likely does not conform to the common mode for RNase-S and pro-OT/BNPI.

of anti-sense binding to immobilized sense peptide by frontal elution analytical affinity chromatography shows that binding is not 1:1, but rather about 5 or 6:1 (at least). Again, the stoichiometry results are more consistent with a model involving sequence matchup along elongated peptides than with recognition of compactly folded peptides. The anti-sense results thus do not fit in any simple way to the orthodox view of macromolecular recognition (Fig. 12). Nonetheless, it is reasonable to suspect that the forces involved in anti-sense/sense recognition are similar to those in more natural peptide and protein recognition events. Describing anti-sense/sense binding mechanisms thus may well help to understand recognition mechanisms in biological peptide and protein systems. In addition, anti-sense peptides may provide useful starting points for the design of recognition molecules for biotechnological applications, although the road to specific recognition molecules may require substantial refinement of antisense sequences per se. The results of synthetic modeling and mutation of anti-sense peptide, currently under investigation in our laboratory (39), hopefully will help to evaluate more fully the nature of the sequence code responsible for sense/anti-sense peptide recognition.

V. CONCLUDING COMMENTS

Synthetic sequence design has proven to be a useful tool to characterize and re-engineer peptide and protein recognition events in biology. The orthodox view of recognition, reflected by sequence simplification studies of RNase-S, can be quite helpful in categorizing and designing sequence information in less well-characterized systems, as discussed here for the oxytocin/neurophysin biosynthetic precursor. Synthesis also has been found to be a valuable means to explore peptide recognition phenomena for de novo-designed "in-vitro" systems. The results with anti-sense peptide suggest that the orthodox view of peptide and protein recognition may be only a variation on a more generalized theme of recognition. Correlating recognition mechanisms of anti-sense peptides with those of more conventional systems such as S-Peptide/S-Protein may provide useful insights into common forces which drive all peptide and protein recognition phenomena.

REFERENCES

1. Ackers, G.K., and Smith, F.R., Annu. Rev. Biochem. 54, 597-629 (1985).
2. Fersht, A.R., Leatherbarrow, R.J., and Wells, T.N.C., Trends in Biochem. Sci. 11, 321-325 (1986).
3. Chaiken, I.M., CRC Crit. Rev. Biochem. 11, 255-301 (1981).
4. Gross, E., and Meienhofer, J., Eds. The Peptides Vol. 2 (Academic Press, New York), 1981.
5. Richards, F.M, and Wyckoff, H.W., in The Enzymes, P.D. Boyer, Ed. (Academic Press, New York), Vol. 4, pp. 647-806 (1971).
6. Finn, F.M., and Hoffman, K., Accts. Chem. Res. 6, 169-176 (1973).
7. Scoffone, E., Marchiori, F., Moroder, L., Rocchi, R., and Borin, G., in Medicinal Chemistry III, Milan, Special Contributions, Pratesi, P., Ed. (Butterworths, London) pp. 83-104, 1973.
8. Komoriya, A., and Chaiken, I.M., J. Biol. Chem. 257, 2599-2604 (1982).
9. Kanmera, T., Homandberg, G.A., Komoriya, A., and Chaiken, I.M., Internat. J. Pept. Prot. Res. 21, 74-83 (1983).
10. Liang, X., Shai, Y., Dalzoppo, D., Fontana, A., and Chaiken, I.M., in preparation.
11. Shai, Y., Flashner, M., and Chaiken, I.M., Biochemistry 26, 669-675 (1987).
12. Taylor, H.C., and Chaiken, I.M., Proc. Natl. Acad. Sci. USA 82, 6423-6426 (1985).
13. Loh, Y.P., Brownstein, M.J., and Gainer, H., Annu. Rev. Neurosci. 7, 189-222 (1985).
14. Chaiken, I.M., Kanmera, T., and Sequeira, R.P., in Opioid Peptides: Molecular Pharmacology, Biosynthesis, and Analysis, Rapaka, R.S., and Hawks, R.L., Eds. (NIDA Press, Rockville) pp. 3-19 (1986).
15. Russell, J.T., Curr. Top. Memb. Res., in press.
16. Land, H., Schutz, G., Schmale, H., and Richter, D., Nature 295, 299-303 (1982).
17. Land, H., Grez, M., Ruppert, S., Schmale, H., Rehbein, M., Richter, D., and Schutz, G., Nature 302, 342-344 (1983).
18. Ando, S., McPhie, P., and Chaiken, I.M., J. Biol. Chem. 262 (1987), in press.
19. Cohen, P., Nicolas, P., and Camier, P., Curr. Top. Cell Regul. 15, 163-318 (1979).

20. Breslow, E., Annu. Rev. Biochem. 48, 251-274 (1979).
21. Chaiken, I.M., Abercrombie, D.M., Kanmera, T., and Sequeira, R.P., in Peptide and Protein Reviews, Vol. 1 (Hearn, M.T.W., Ed., Marcel Dekker, New York), pp. 139-209 (1983).
22. Pickering, B.T. and Swann, R.W., in The Cell Biology of the Secretory Process (Cantin, M., Ed.,S. Karger, Berlin), pp. 247-275 (1984).
23. Breslow, E., Ibid, pp. 276-308 (1984).
24. Chaiken, I.M., Tamaoki, H., Brownstein, M.J., Gainer, H., FEBS Lett. 164, 361-365 (1983).
25. Kanmera, T., and Chaiken, I.M., J. Biol. Chem. 260, 8474-8482 (1985).
26. Chaiken, I.M., Randolph, R.E., and Taylor, H., Ann. N.Y. Acad. Sci. 248, 442-450 (1975).
27. Menendez-Botet, C.J., and Breslow, E., Biochemistry 14, 3825-3835 (1975).
28. Angal, S., and Chaiken, I.M., Biochemistry 21, 1574-1580 (1982).
29. Chaiken, I.M., J. Chromatogr. Biomed. Applications 376, 1-32 (1986).
30. Breslow, E., Proc. Natl. Acad. Sci. USA 67, 493-500 (1970).
31. Fassina, G., and Chaiken, I.M., unpublished data.
32. Ando, S., Murthy, A.S.N., Eipper, B.A., and Chaiken, I.M., unpublished data.
33. Bost, K.L., Smith, E.M., and Blalock, J.E., Proc. Natl. Acad. Sci. USA 82, 1372-1375 (1985).
34. Bost, K.L., and Blalock, J.E., Mol. Cell. Endocrinol. 44, 1-9 (1986).
35. Bost, K.L., Smith, E.M., and Blalock, J.E., Biochem. Biophys. Res. Commun. 128, 1373-1380 (1985).
36. Swaisgood,H.E. and Chaiken, I.M., in Analytical Affinity Chromatography (Chaiken, I.M., Ed., CRC Press, Boca Raton, 1987), in press.
37. Fassina, G., and Chaiken, I.M., in Adv. Chromatogr. (1987), in press.
38. Chiancone, E., and Gattoni, M., this volume, pp. 321-332.
39. Shai, Y., and Chaiken, I.M., Fed. Proc. 46, 2022 (1987).

LIMITED PROTEOLYSIS OF GLOBULAR PROTEINS: MOLECULAR ASPECTS DEDUCED FROM STUDIES ON THERMOLYSIN

Claudio Vita, Daniele Dalzoppo and Angelo Fontana

Department of Organic Chemistry, Biopolymer Research Centre of CNR, University of Padua, 35131 Padua, Italy

The term "limited proteolysis" of proteins was coined to indicate the specific fission of only a few peptide bonds among the many present in a protein molecule substrate (1). Limited proteolysis plays a fundamental role in many biological processes, such as zymogen activation, hormone production from protein precursor, intracellular protein turnover, post-translational processing of proteins, transport across a membrane, blood coagulation, fertilization, supramolecular assembly, etc. (2-6). Thus, elucidation of the mechanism of biorecognition between the enzyme/protease and its substrate/protein (usually globular) leading to restricted and specific fissions of protein chains would be of paramount importance for an understanding of the molecular aspects of a multitude of physiological processes.

Limited proteolysis has proved to be a useful

Abbreviations used: EDTA, ethylenediaminetetraacetic acid; RP-HPLC, reverse phase high performance liquid chromatography; SDS-PAGE, polyacrylamide gel electrophoresis in the presence of 0.1% sodium dodecyl sulfate; Tris, tris-(hydroxymethyl)-aminomethane.

technique in protein chemistry studies in order to produce from globular proteins and enzymes quite numerous "nicked" protein species of functional and structural interest. Nicked proteins are given when protein fragments remain associated in relatively stable and often functional complexes (6). The most classical example is the proteolysis of ribonuclease A by subtilisin to produce the stable, active complex ribonuclease S, given by a tight association of S peptide and S protein (7). Moreover, the technique of limited proteolysis has been used to elucidate the domain-like structure of relatively large globular proteins, allowing isolation of fragments capable of independent folding (8).

The specificity of limited proteolysis of globular proteins could be explained in terms of a protein-protein interaction process in which only a few chain segments of the globular protein substrate are characterized by accessibility and adaptability to the active site of the attacking protease. Actually, this is seldom achieved with many small globular proteins, which are quite resistant to proteolysis in their native state, but easily degraded in their unfolded state. This lead to the idea that the digestion of globular proteins proceeds *via* the unfolded state only, leading to many small molecular weight peptides if a protease of low specificity is employed (9-12).

In this communication we wish to summarize the results of our studies aimed at characterizing the pattern of limited proteolysis of thermolysin by added protease or by autolysis under different experimental conditions (13-16). Thermolysin is a highly thermostable metalloendoprotease obtained from *B.thermoproteolyticus* (17) for which the amino acid sequence of its chain of 316 amino acid residues (18) and three-dimensional structure (19-20) are known. Recently, a highly refined structure of thermolysin at 1.6 Å resolution has been reported, together with the average thermal motion of its polypeptide chain (B factor profile) (21). Thus, thermolysin can be listed as one of the enzyme molecules that has been characterized in great detail, making this protein an interesting model for studying structure-function relationships as well as folding and stability properties of proteins.

The results of our studies (13-16) indicate that exposed and flexible loops of the thermolysin molecule are the most susceptible sites of limited proteolysis (15). These results are compared with data of limited proteolysis of globular proteins reported in the literature and some general molecular aspects of the limited proteolysis process are discussed.

LIMITED PROTEOLYSIS AND AUTOLYSIS OF THERMOLYSIN

Thermolysin was subjected to limited proteolysis by subtilisin Carlsberg or to autolysis upon incubation at 55°C or in the presence of the metal-chelating agent EDTA (1-10 mM). The actual experimental conditions of limited proteolysis are reported in the legend to Fig. 1 (see also previous publications from our laboratory, 13-16). The SDS-PAGE analysis (22) of the proteolytic mixtures revealed that the fragmentation reactions occuring under the specified conditions are quite selective and limited, since only a few protein fragments of relatively high molecular weight (ranging from 23,000 to 9,000) are seen in the gels.

When the proteolytic digests were individually loaded on a gel filtration column equilibrated and eluted with an aqueous buffer at neutral pH, the fragments were eluted together from the column and moreover with the same elution volume as intact thermolysin, indicating that the proteolytic processes produce "nicked" protein species. A combination of gel filtration and affinity chromatography permitted the isolation of two- as well as three-fragment complexes in homogeneous form and in high yields (70-80 %) (13-16).

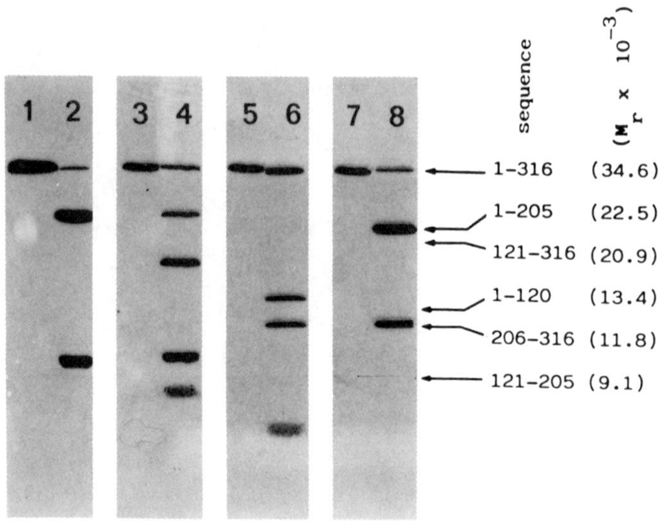

Figure 1. Limited proteolysis or autolysis of thermolysin (Sigma, St. Louis, Mo, USA) monitored by SDS-PAGE of digestion products. A vertical slab gel with an exponential gradient from 15 to 24 % along the direction of migration was used (22). The gel (1 mm thickness) was stained with Coomassie Brilliant Blue R-250. Samples (10 µl containing 5 µg of protein) were taken from the reaction mixtures, mixed with o-phenanthroline (5 µl of a 0.1 M solution in water) to inhibit thermolytic activity (13) and then mixed with the sample buffer of the electrophoresis system. Lane 1,3,5,7: Thermolysin dissolved (1 mg/ml) in 20 mM Tris-HCl buffer, pH 7.2, containing 10 mM $CaCl_2$, and incubated for 24 hrs at room temperature. Lane 2: Limited proteolysis (24 hrs, 37°C) of thermolysin dissolved (0.8 mg/ml) in 50 mM Tris-HCl buffer, pH 9.0, containing 10 mM $CaCl_2$ and 5 mM o-phenanthroline, by subtilisin Carlsberg (Sigma) (1:50 molar ratio by weight). Lane 4: Thermal autolysis of thermolysin (60 hrs, 55°C) dissolved (0.6 mg/ml) in 50 mM Tris-HCl buffer, pH 9.0, containing 10 mM $CaCl_2$. Lane 6: Autolysis of thermolysin (2 days at room temperature) dissolved (0.6 mg/ml) in 10 mM Tris-HCl buffer, pH 7.2,

Figure 1. (continued) containing 0.1 M NaCl, 1.5 mM $CaCl_2$ and 1 mM EDTA. Lane 8: Autolysis of thermolysin (20 hrs at room temperature) dissolved (0.7 mg/ml) in 10 mM Tris-HCl buffer, pH 7.2, containing 0.1 M NaCl, 1.5 mM $CaCl_2$ and 10 mM EDTA. In the right part of the figure are indicated the peptides sequences and the molecular weights of the cyanogen bromide fragments of thermolysin used as molecular weight markers and obtained by partial chemical cleavage (100 equivalents of reagent per mole of protein, 50 hrs at room temperature in 70 % aqueous formic acid) at the level of the methionine residues in position 120 and 205 of the polypeptide chain of 316 amino acid residues of thermolysin (18).

The nicked thermolysin species were folded into a stable structure similar to that of the intact protein, as indicated by far-ultraviolet circular dichroism measurements and by the fact that all the nicked species are recognized and precipitated by rabbit anti-thermolysin antibodies (not shown). Apparently, all these species are devoid of enzymatic activity, with the exception of the two-fragment complex produced by the subtilisin cleavage. This species, called thermolysin S, shows reduced but intrinsic activity (~3 % of that of the native thermolysin) (13).

When gel filtration chromatography of the proteolytic digests was performed in the presence of 6 M guanidine hydrochloride, the thermolysin fragments were eluted from the column according to their size and then recovered after dialysis and lyophilization. These fragments were also analyzed and purified by RP-HPLC using analytical and semipreparative μBondapak C-18 columns eluted with an aqueous acetonitrile gradient containing 0.1 % trifluoroacetic acid (14-16). A number of protein fragments were thus isolated to homogeneity and their identity established by amino acid analysis after acid hydrolysis (6 N HCl, 110°C, 22 hrs), determination of the

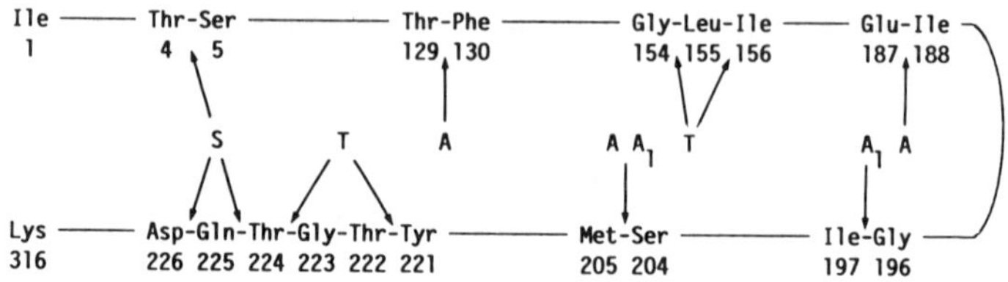

Figure 2. Schematic representation of the polypeptide chain of thermolysin showing the bonds cleaved by limited proteolysis or autolysis. S, site of cleavage by subtilisin; A, cleavage by autolysis in the presence of 1.5 mM $CaCl_2$ and 1 mM EDTA; A1 , cleavage by autolysis in the presence of 1.5 mM $CaCl_2$ and 10 mM EDTA; T, cleavage by thermal autolysis (actual experimental conditions described in the legend to Fig. 1).

NH_2 -terminal sequences after three steps of Edman degradation (23), as well as COOH-terminal sequences by determining the time course of the release of amino acids during carboxypeptidase Y digestion (15). Comparison of all these data with the known sequence of thermolysin allowed unambigous identification of the protein fragments isolated and thus of the bonds cleaved along the polypeptide chain (cf. Fig. 2) under the specific conditions of proteolysis indicated in the legend to Fig. 1.

CORRELATION OF CLEAVAGE SITES WITH STRUCTURE AND DYNAMICS OF THE THERMOLYSIN MOLECULE

The selectivity with which limited proteolysis occurs in the thermolysin molecule, as indicated by the small number of fragments generated, appears to be quite striking. Considering the broad specificity of both subtilisin (24) and thermolysin (25), clearly conformational features of the globular protein substrate must dictate the sites of attack on the protein surface.

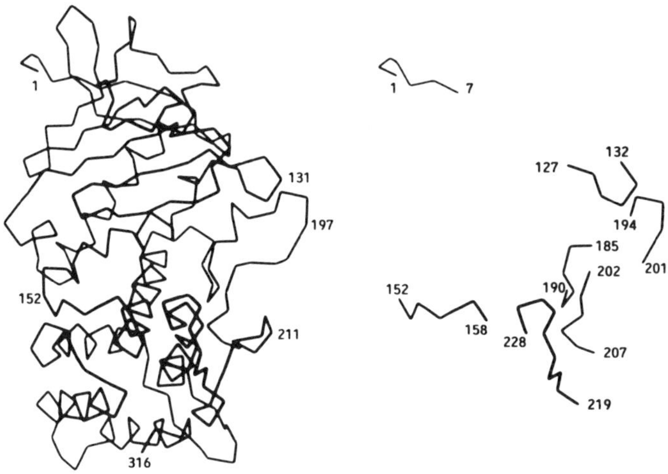

Figure 3. Computer generated models of the α-carbon backbone of thermolysin (left) and of the chain segments (right) supposed to interact with the active sites of subtilisin or thermolysin in the limited proteolysis and autolysis processes herewith described.

Fig. 3 shows (left) the polypeptide chain of thermolysin and (right) a set of six-residues long chain segments comprising three residues both at the NH_2 - as well as COOH-side of every bond actually cleaved (cf. Fig. 2). These stretches of six amino acid residues may represent the portion of the polypeptide chain on the thermolysin surface interacting with the protease active site in the various processes of proteolysis described. In fact, it has been shown that a segment of at least six residues has to interact with the protease active site in order to obtain an effective interaction leading to peptide bond cleavage (26-28).

Fig. 3 provides evidence that the observed sites of cleavage are located at exposed bends or loops of the thermolysin polypeptide chain, never within segments of ordered secondary structure. For example, thermal autolysis leads to fission at peptide bonds 154-155 and 155-156, located on a loop connecting two helical segments, and also at loop 220-226. Subtilisin also cleaves at this loop, as well as at the amino terminus. In the presence of EDTA, which removes looser calcium ion(s) from the metallo-protein (15,29), fission occurs at the peptide bond 129-130, located on a surface turn, and at a long surface loop 180-210, held in place by calcium ions in the native molecule (30). Thus, exposed loops of the thermolysin molecule seem to be the loci on the protein structure more easily recognized by the protease active site.

The next problem to examine is why only those loops of the globular protein substrate shown in Fig. 3, among all those present at the protein surface, are selected by the attacking protease. In this respect, it is conceivable to suggest that limited proteolysis of a native globular protein would occur at loops of the polypeptide chain which are not only exposed, but also quite flexible, since this would facilitate the most advantageous adaptation of the polypeptide substrate to the stereochemistry of the active site of the protease (3,6,31). In fact, a native globular protein is not a static entity, but rather one which undergoes conformational fluctuations about its most stable conformation, as clearly emphasized in recent years by numerous studies using quite a number of physical methods (32-36). More clear-cut evidence for motions in proteins comes directly from the highly refined analysis of the diffraction data giving information about mean atom displacements, as usually expressed by the crystallographically determined temperature factors (B values) (37,38). These factors when plotted against the residue number, provide a graphic image of the degree of mobility existing along the polypeptide chain.

Fig. 4 shows the profile of the crystallographic mean-square displacement of the α-carbon atoms along the polypeptide chain of thermolysin (21). In the figure the regions of ordered secondary structure (α-helices and

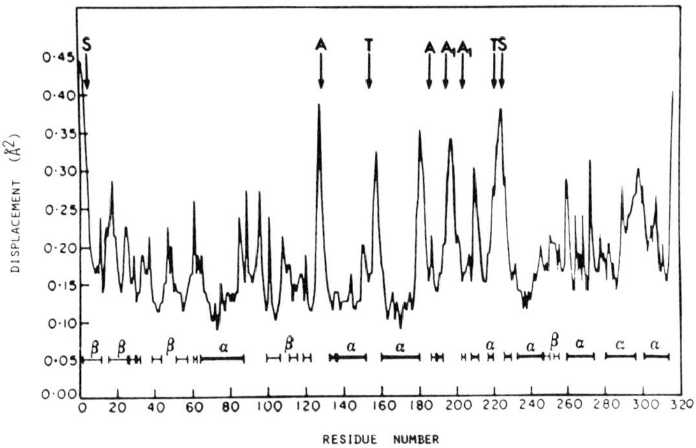

Figure 4. Plot of the average main-chain temperature factors along the polypeptide chain of thermolysin (taken from ref. 14 and 21, with permission). Bars at the bottom of the figure indicate segments of secondary structure (helices and strands). Arrows indicate sites of limited proteolysis or autolysis of thermolysin under the different experimental conditions reported in the legend to Fig. 1. Symbols employed to indicate sites of cleavage are those indicated in the legend to Fig. 2.

sheets) are also shown, as well as all the sites of limited proteolysis. The correlation between cleavage sites and maxima of flexibility (segmental mobility) of the thermolysin polypeptide chain appears to be quite striking. Nearly all the maxima correspond within a few residues to the sites of protease attack. In addition, it should be noted that high mobility as well as cleavage sites do not occur within segments of regular secondary structure, but only at loops or turns. Taking together the results shown in Fig. 3 and 4, it can be concluded that the sites of proteolytic attack on the thermolysin molecule involve exposed loops characterized by flexibility.

LIMITED PROTEOLYSIS OF OTHER PROTEIN MOLECULES

Among the numerous examples of limited proteolysis of globular proteins reported in the literature, we will herewith consider those involving proteins of known three-dimensional structure. Ribonuclease A is cleaved by subtilisin in the region 20-22, giving ribonuclease S (7,39); elastase (40) and chymotrypsin (41) cleave essentially at the same region (18-21 and 25-26, respectively). Trypsin and ficin cleave mainly in the 31-34 region (42,43), with ficin cleaving also at the 104-105 bond (43). All these sites in the ribonuclease molecule are located at exposed loops, outside segments of secondary structure, and moreover in regions associated with high B factors (44). Staphylococcal nuclease is cleaved by trypsin at bonds 5-6, 48-49 and 49-50, giving the two-fragment complex nuclease-T (45,46). In the electron density map of nuclease all these regions are either invisible or of low density, indicating flexibility (47). The "activation sites" of trypsinogen (segments 1-6, 133-144, 175-184 and 208-214) are also invisible in the electron density map of the protein (48); these sites are readily cleaved by proteases. In the case of elastase, trypsin cleaves selectively bond 125-126 (49), located in an exposed and flexible loop in the protein.

Several other examples of limited proteolysis occurring at flexible "hinges", poorly resolved in the electron density map, have been reported in the literature but are not discussed in detail here. Some of these examples include immunoglobulin, phosphorylase, citrate synthase, transferrin, λ-repressor and calmodulin.

DISCUSSION

The results of this study show that limited

proteolysis of the thermolysin molecule under a variety of experimental conditions occurs at exposed and flexible loops, thus complementing previous studies and proposals that sites of limited proteolytic cleavage of globular proteins are expected to occur at surface, random loops of the protein, and not at internal loci or within segments of regular secondary structure of the protein molecule (6). This study emphasizes the role of both loops and chain flexibility in recognition processes of globular proteins, and thus these characteristics of the interacting molecular surfaces should be considered together with other properties such as complementarity, hydrophobicity and electrostatic potential (50,51). The fact that chain loops are involved in the proteolysis event is in keeping with the proposed "functional" role of loops in proteins, these being involved in several fundamental protein recognition processes, such as phosphorylation, glycosylation, ion binding, as well as antigenicity (52,53).

In other studies of protein-protein interactions emphasis has been given to other characteristics of the interacting protein surfaces, such as hydrophilicity (52,54), turns or loops (53) or surface accessibility (55,56). It is clearly demonstrated that turns (bends or loops) hydrophilicity-segmental mobility-surface accessibility are all surface characteristics of globular proteins and that all are strongly correlated (57-58). Thus, it appears that individual authors probably are prone to emphasize one or another of these surface properties depending upon the actual experiments performed or the protein system under study. In a recent study (59), it was claimed that likely protein accessibility rather than mobility determines the sites of limited proteolysis. However, this conclusion does not seem properly supported by the data shown, since the only case of limited proteolysis occurring at a "rigid" but accessible site in bovine trypsinogen (bond Arg 105-Val 106) refers to the tryptic autolytic reaction in commercial samples of bovine

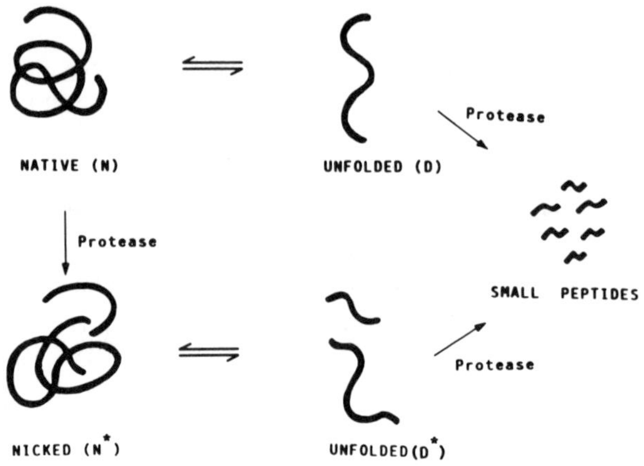

Figure 5. Proposed general scheme of the proteolytic degradation process of a globular protein. N, native protein; D, unfolded protein; N*, nicked protein; D*, unfolded nicked protein.

trypsinogen. Actually, this bond is not cleaved in porcine trypsin, whereas selective and essentially quantitative cleavage is observed at peptide bond Lys 131-Ser 132 only (60). In addition, it should be considered that data of limited proteolysis can be interpreted in terms of conformational (accessibility, flexibility) parameters only if proteases of low specificity are employed; trypsin cleaves at Lys and Arg residues only.

The scheme depicted in Fig. 5 helps to rationalize the data of the proteolysis processes of thermolysin here reported, as well as those obtained with other protein systems (61,62). Digestion of globular proteins often has been interpreted in terms of a conformational transition between the native (N) and the unfolded state (D), implying that state D is the only true substrate for the attacking protease (9-12). This two-state model implies an all-or-none digestion of the globular protein, leading to the formation of a large number of small peptides as a result of the presence of many susceptible sites in the unfolded state (D) of the protein. However, the results of this study and others reported in the literature (61,62) indicate that the initial steps of the proteolysis event involve scission of very few peptide bonds leading to the formation of nicked protein species (N*) still maintaining a folded three-dimensional structure. The nicked species is much less stable than the native one (N) and unfolds more readily to the denatured state (D*), which in turn is degraded much faster by the protease to small peptides. As a matter of fact, all the nicked species of thermolysin here described, isolated to homogeneity by chromatography (13-16), are degraded by the action of added protease (trypsin, chymotrypsin) much faster than native thermolysin (unpublished results). The apparent all-or-none nature of the proteolysis process often observed with many globular proteins (9-12) can be explained by a slow initiation of the first cleavage and successive rapid digestion. Usually, nicked protein species, contrary to the thermolysin case, are very labile molecules and their easy unfolding is followed by very rapid digestion.

CONCLUSION

A result of particular interest in this study is that it shows that proteolytic enzymes can be used to pinpoint, in globular proteins, sites (loops or turns) characterized by local flexibility that likely could be difficult or impossible to detect by other physical methods. In this respect, the enzyme-probe method would complement methods currently employed to predict segments of high antigenicity based on hydrophilicity (52,54), reverse-turn potential (53), surface accessibility (55) or segmental mobility (63-65). All these last methods, used in conjunction with the limited proteolysis approach, could facilitate selection of chain segments suitable for inducing antibodies which cross-react with native proteins.

REFERENCES

1. Linderstrom-Lang, K. and Ottesen, M. (1947) Comp. Rend. Trav. Lab. Carlsberg 26, 403-442.
2. Goldberg, A.L. and Dice, J.F. (1974) Ann. Rev. Biochem. 43, 835-869.
3. Neurath, H. and Walsh, K.A. (1976) Proc. Natl. Acad. Sci. USA 73, 3825-3832.
4. Ribbons, D.W. and Brew, K., eds. (1976) Proteolysis and Physiological Regulation, Academic Press, New York.
5. Neurath, H. and Walsh, K.A. (1976) in: Proteolysis and Physiological Regulation (Ribbons, D.W. and Brew, K., eds.), pp. 29-42, Academic Press, New York.
6. Neurath, H. (1980) in: Protein Folding (Jaenicke, R., ed.) pp. 501-504, Elsevier/North Holland Biomedical Press, Amsterdam and New York.
7. Richards, F.M. and Vithayathil, P.J. (1959) J. Biol. Chem. 234, 1459-1464.
8. Wetlaufer, D.B. (1981) Adv. Prot. Chem. 34, 61-92.
9. Matthyssens, G.E., Simons, G. and Kanarex, L. (1972) Eur. J. Biochem. 26, 449-454.
10. McLendon, G. and Radany, E. (1977) Biochem. Biophys. Res. Commun. 77, 959-966.
11. Mc Lendon, G. and Radany, E. (1978) J. Biol. Chem. 253, 6335-6337.
12. Imoto, T., Yamada, H. and Ueda, T. (1986) J. Mol. Biol. 190, 647-649.
13. Vita, C., Dalzoppo, D. and Fontana, A. (1985) Biochemistry 24, 1798-1806.
14. Fontana, A., Fassina, G., Vita, C., Dalzoppo, D., Zamai, M. and Zambonin, M. (1986) Biochemistry 25, 1847-1851.
15. Fassina, G., Vita, C., Dalzoppo, D., Zamai, M., Zambonin, M. and Fontana, A. (1986) Eur. J. Biochem. 156, 221-228.

16. Fontana, A., Fassina, G., Vita, C., Dalzoppo, D., Zamai, M. and Zambonin, M. (1986) in: Zinc Enzymes (Bertini, I., Luchinat, C., Maret, W. and Zeppezauer, M., eds.) pp.225-237, Birkhäuser Verlag, Basel.
17. Endo, S. (1962) J. Ferment. Technol. 40, 346-353.
18. Titani, K. Hermodson, M.A., Ericsson, L.H., Walsh, K.A. and Neurath, H. (1972) Biochemistry 11, 2427-2435.
19. Colman, P.M., Jansonius, J.N. and Matthews, B.W. (1972) J. Mol. Biol. 70, 701-724.
20. Matthews, B.W., Weaver, L.H. and Kester, W.A. (1974) J. Biol. Chem. 249, 8033-8044.
21. Holmes, M.A. and Matthews, B.W. (1982) J. Mol. Biol. 160, 623-639.
22. Laemmli, U.K. (1970) Nature (London) 227, 680-685.
23. Tarr, G.E. (1982) in: Methods in Protein Sequence Analysis (Elzinga, M., ed.) pp. 223-232, Humana Press, Clifton, N.J., USA.
24. Markland, F.S.Jr. and Smith, E.L. (1971) Methods Enzymol. 3, 561-608.
25. Heinrikson, R.L. (1977) Methods Enzymol. 47, 175-189.
26. Ottesen, M. (1967) Ann. Rev. Biochem. 36, 55-76.
27. Berger, A. and Schechter, I. (1969) Phil. Trans Royal Soc. (London) 257, 249-264.
28. Morihara, K. and Tsuzuki, M. (1970) Eur. J. Biochem. 15, 374-380.
29. Weaver, L.H., Kester, W.R., Ten Eyck, L.F. and Matthews, B.W. (1976) in: Enzymes and Proteins from Thermophilic Microorganisms (Zuber, H., ed.), Birkhäuser Verlag, Basel, Experientia Suppl. 26, 31-39.
30. Roche, R.S. and Woordouw, G. (1978) CRC Crit. Rev. Biochem. 5, 1-23.
31. Mihalyi, E. (1978) Application of Proteolytic Enzymes to Protein Structure Studies, Vol.1, CRC Press Inc., West Palm Beach, USA.
32. Wüthrich, K. and Wagner, G. (1978) Trends Biochem. Sci. 3, 227-230.
33. Gurd, F.R.N. and Rothgeb, T.M. (1979) Adv. Protein Chem. 33, 73-165.

34. Karplus, M. and McCammon, J.A. (1983) Ann. Rev. Biochem. 53, 263-300.
35. Ringe, D. and Petsko, G.A. (1985) Prog. Biophys. Molec. Biol. 45, 197-235.
36. Huber, R. (1979) Trends Biochem. Sci. 4, 271-276.
37. Frauenfelder, H., Petsko, G.A. and Tsernoglou, D. (1979) Nature (London) 280, 558-563.
38. Artymiuk, P.J., Blake, C.C.F., Grace, D.E.P., Oatley, S.J., Phillips, D.C. and Sternberg, M.J.E. (1979) Nature (London) 280, 563-568.
39. Doscher, M.S. and Hirs, C.H.W. (1967) Biochemistry 6, 304-312.
40. Klee, W.A. (1965) J.Biol.Chem. 240, 2900-2906.
41. Rupley, J.A. and Scheraga, H.A. (1963) Biochemistry 2, 421-431.
42. Ooi, T., Rupley, J.A. and Scheraga, H.A. (1963) Biochemistry 2, 432-437.
43. Winchester, B.G., Mathias, A.P. and Rabin, B.R. (1970) Biochem. J. 117, 299-307.
44. Wlodawer, A., Bott, R. and Sjolin, L. (1982) J. Biol. Chem. 257, 1325-1332.
45. Taniuchi, H., Anfinsen, C.B. and Sodja, A. (1967) Proc. Natl. Acad. Sci. USA 58, 1235-1242.
46. Taniuchi, H. and Anfinsen, C.B. (1968) J. Biol. Chem. 243, 4778-4786.
47. Arnone, A., Bier, C.J., Cotton, F.A., Day, V.W., Hazen, E.E. Jr., Richardson, D.C., Richardson, J.S. and Yonath, A. (1971) J. Biol. Chem. 246, 2302-2316.
48. Bode, W., Fehlhammer, H. and Huber, R. (1976) J. Mol. Biol. 106, 325-335.
49. Ghelis, C., Tempete-Gaillourdet, M. and Yon, J.M. (1978) Biochem. Biophys. Res. Commun. 84, 31-36.
50. Blundell, T.L. (1981) in: Structural Aspects of Recognition and Assembly in Biological Macromolecules (Balaban, M., Sussman, J.L., Traub, W., and Yonath, A., eds.) pp. 281-286, Balaban International Sciences Services, Philadelphia, PA.
51. Katchalski-Katzir, E. (1983) in: Affinity Chromatography and Biological Recognition (Chaiken,

I.M., Wilchek, M. and Parikh, I., eds.) pp. 7-26, Academic Press, New York.

52. Hopp, T.P. and Woods, K.R. (1981) Proc. Natl. Acad. Sci. USA 78, 3824-3828.
53. Rose, G.D., Gierasch, L.M. and Smith, J.A. (1985) Adv. Protein Chem. 37, 1-100.
54. Kyte, J. and Doolittle, R.F. (1982) J. Mol. Biol. 157, 105-132.
55. Novotny, J., Handschumacher, M., Haber, E., Bruccoleri, R.E., Carlson, W.B., Fanning, D.W., Smith, J.A. and Rose, G.D. (1986) Proc. Natl. Acad. Sci. USA 83, 226-230.
56. Fanning, D.W., Smith, J.A. and Rose, G.D. (1986) Biopolymers 25, 863-883.
57. Bennett, W.S. and Huber, R. (1984) CRC Crit. Rev. Biochem. 15, 291-384.
58. Petsko, G.A. and Ringe, D. (1984) Ann. Rev. Biophys. Bioeng. 13, 331-372.
59. Novotny, J. and Bruccoleri, E. (1987) FEBS Lett. 211, 185-189.
60. Hermodson, M.A., Ericsson, L.H., Neurath, H. and Walsh, K.A. (1973) Biochemistry 12, 3146-3153.
61. Endo, S., Nagayama, K. and Wada, A. (1985) J. Biomol. Struct. Dynamics 3, 409-421.
62. Ogasahara, K., Tsunazawa, S., Soda, Y., Yutani, K. and Sugino, Y. (1985) Eur. J. Biochem. 150, 17-21.
63. Westhof, E., Altschuh, D., Moras, D., Bloomer, A.C., Mondragon, A., Klug, A. and Van Regenmortel, M.H.V. (1984) Nature (London) 311, 123-126.
64. Tainer, J.A., Getzoff, E.D., Alexander, H., Houghten, R.A., Olson, A.J., Lerner, R.E. and Hendrickson, W. A. (1984) Nature (London) 311, 127-134.
65. Tainer, J.A., Getzoff, E.D., Paterson, Y., Olson, A.J. and Lerner, R.A. (1985) Ann. Rev. Immunol. 3, 501-535.

CRYSTALLOGRAPHIC STUDIES ON RETINOL-BINDING PROTEIN AND BETA LACTOGLOBULIN

Hugo L. Monaco, Giuseppe Zanotti and
Paola Spadon
Biopolymer Research Centre of CNR, Department of Organic Chemistry, University of Padua - 35131 Padua, Italy

It is not difficult to understand the need for a carrier for retinol in plasma since the molecule is not soluble in an aqueous medium. The protein that effects this function is called plasma Retinol-Binding Protein (RBP) (for a recent review see reference 1). Two other additional functions are performed by this protein: protection of the labile retinol molecule and selective delivery, since the vitamin is only transferred from the carrier to the cells that have a specific membrane receptor (2,3). Retinol-Binding Proteins have been characterized in the plasma of many species including man. The retinol transport system in human plasma includes another well known protein, transthyretin, formerly known as prealbumin. The two proteins form, under physiological conditions, a one to one complex that has an enhanced tendency to dissociate upon delivery of the vitamin to the receptors. Other conditions that favor dissociation are low ionic strength (4) and perhaps disruption of hydrophobic contacts (5).

Human plasma RBP is a single polypeptide chain of 182 aminoacids with three disulphide bridges and a single high affinity binding site for retinol (6). Its three-

Table 1: The trigonal crystals of RBP

Space group	R3
a	104.2 Å
b	104.2 Å
c	74.5 Å
V	699,042 $Å^3$
V_m	1.87
n	2

dimensional structure has been under investigation for some years in our laboratory and at the University of Upsala in Sweden. A preliminary report on the chain trace based on 3 Å data of an orthorhombic crystal form has been published by the Uppsala group (7). We have been working on a trigonal crystal form which contains two molecules in the asymmetric unit. Table 1 gives the unit cell parameters and other characteristics of these crystals.

A large conformational change has been invoked to explain the reduced affinity of apo-RBP for transthyretin (7). Figure 1 shows pictures of our crystals of apo and holo-RBP.

The main difference between the two crystal types is the presence or absence of fluorescence which is obviously due to the retinol molecule. In spite of the fact that the habit is often different, the two crystals are isomorphous as we have reported elsewhere (8). The isomorphism of apo and holo-RBP gives a clear indication that no drastic conformational change exists in vitro between holo and artificially produced apo-RBP. We have used two methods to generate apo from holo-RBP: organic solvent extraction and hydrophobic interaction chromatography. In both cases the result is the same. If the apoprotein that has interacted with the receptors is not si-

Figure 1 - Trigonal crystals of holo (left) and apo (right) RBP.

gnificantly different from artificially produced apo-RBP, a detailed description of the differences observed in the two crystals may suggest a mechanism to explain the reduced affinity of RBP for transthyretin after vitamin delivery.

Using conventional X-ray diffraction methods, we have calculated an electron density map of holo-RBP to a nominal resolution of 2.5 Å. Our phases were calculated using three heavy atom derivatives: $EuCl_3$, K_3IrCl_6 and Mercury Phenylglyoxal. The final M.I.R. figure of merit was .66 for all the data included to 2.5 Å resolution. On the ba-

Figure 2 - The two molecules of RBP present in the asymmetric unit of the trigonal crystals.

sis of this electron density map, independent models for the two molecules present in the crystallographic asymmetric unit were built and then refined using the restrained least squares method of Hendrickson and Konnert (9). After completely rebuilding of the models twice, the conventional crystallographic R factor is 28%. This process has not yet converged and so the apo holo differences will not be discussed here. In this connection it should be mentioned that the crystals of apo-RBP do not diffract as well as those of the holo-counterpart. Using the oscillation method and a GX20 Elliott rotating anode X-ray generator as the source, we have collected an essentially complete data set of apo-RBP to 2.8 Å resolution. Figure 2 shows the two molecules present in the crystallographic asymmetric unit. The two molecules have been refined independently so that if there are any differences, presumably due to the crystal packing, they will become apparent in the final refined map.

The molecule of RBP is made up of a single domain constituted by 8 strands of antiparallel beta structure and a very short segment of alpha helix which runs from aminoacids 147 to 160. The retinol molecule binds in the center of this beta barrel and is completely sorrounded by hydrophobic residues. The RBP molecule is shaped like a calyx sorrounding the retinol molecule which is found with its beta ionone ring deepest in the cavity and the hydroxyl group pointing to the outside. The overall molecular architecture seems to be designed to accomodate a hydrophobic cylindrical shape in the center of the barrel. Figure 3 gives a view of the interactions of the retinol and the protein molecules. We have omitted from the model the aminoacid side chains for the sake of clarity. A more detailed description of the molecular architecture and interactions will be published elsewhere.

Shortly after the discovery of RBP, Futterman and Heller studied the interaction of retinol with several proteins and found that bovine Beta Lactoglobulin (BLG) formed a strongly fluorescent water soluble complex with it (10). This result was not entirely surprising since it was already known that BLG possesses a hydrophobic binding site capable of binding alkanes and aromatic hydrophobes with a very high binding constant (11, 12, 13). The protein BLG had been at the time very extensively studied, a fact mainly due to its ready availability. Discovered and crystallized in 1934 by Palmer (14) it has been used as a model to test different

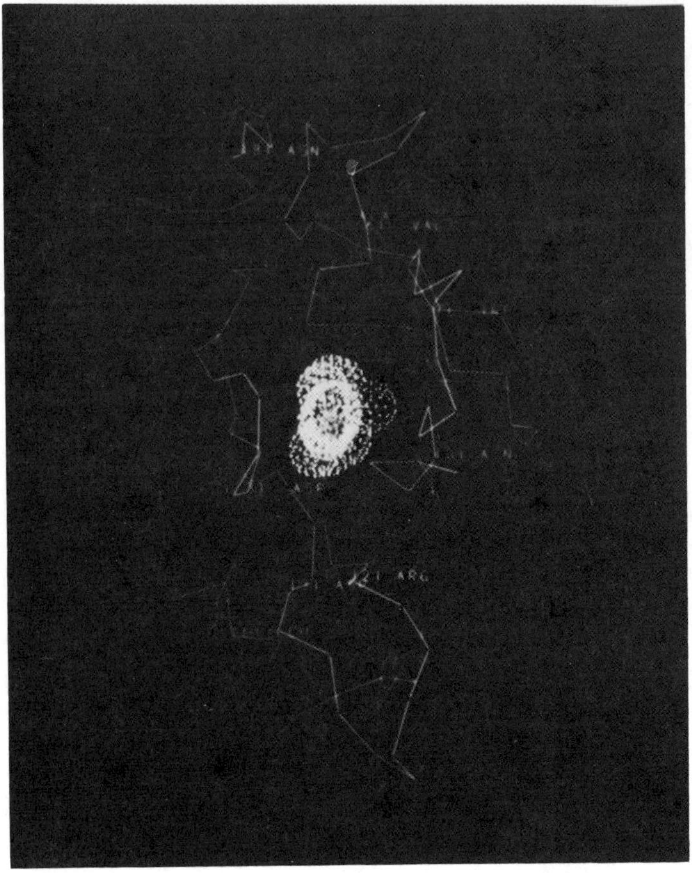

Figure 3 RBP Interaction with retinol

physicochemical techniques in pretty much the same fashion as serum Albumin. In spite of all the information gathered, the function of this molecule remains unknown. More recently, the binding of retinol to BLG has been characterized more accurately (15, 16). The most important findings are that the binding site is one per monomer of BLG, that binding is not pH dependent, that

Table 2: The trigonal crystals of BLG

Space group	$P3_221$
<u>a</u>	54.4 Å
<u>b</u>	54.4 Å
<u>c</u>	113.2 Å
V	690,118 $Å^3$
V_m	2.69
n	1

the site is hydrophobic and contains a Lys nearby (17).

Several crystalline forms of BLG have been studied at low resolution (18) and two have been solved at high resolution. They are an orthorhombic form studied by Sawyer, North and their groups in Edinburgh and Leeds (19) and a trigonal form studied by us in Padova and Pavia. Table 2 gives the unit cell parameters of the BLG trigonal crystals.

Using the multiple isomorphous replacement method, we produced an electron density map of the trigonal crystal form to a nominal resolution of 2.5 Å. Our phases were calculated with three heavy atom derivatives: mercuric acetate, mersalyl, and mercury phenylglyoxal and the overall figure of merit was 0.64. In this electron density map, a short alpha helical region and pieces of beta structure were identifiable. We took the preliminary coordinates of the orthorhombic form solved by Sawyer et al. (19) and rotated the model into the electron density. After manual adjustment of the model to the density, we started to refine using the Hendrickson Konnert method (9). Refinement was performed alternating cycles of reciprocal space restrained refinement and manual

Figure 4 - Polypeptide chain fold of BLG

rebuilding of the model. We rebuilt the model 8 times and performed 144 cycles of refinement. The final model has a conventional crystallographic R factor of 0.22 for data that extend to 2.5 Å resolution.

Though RBP and BLG differ in chain length, BLG being only 162 aminoacids long (20) and in spite of the fact that their sequence homology is limited (21) their three-dimensional structural homology is remarkable (19). BLG is also an eight stranded beta barrel with a short piece of alpha helix which runs from aminoacids 131 to 140. Figure 4 shows the alpha carbons of the BLG model built for

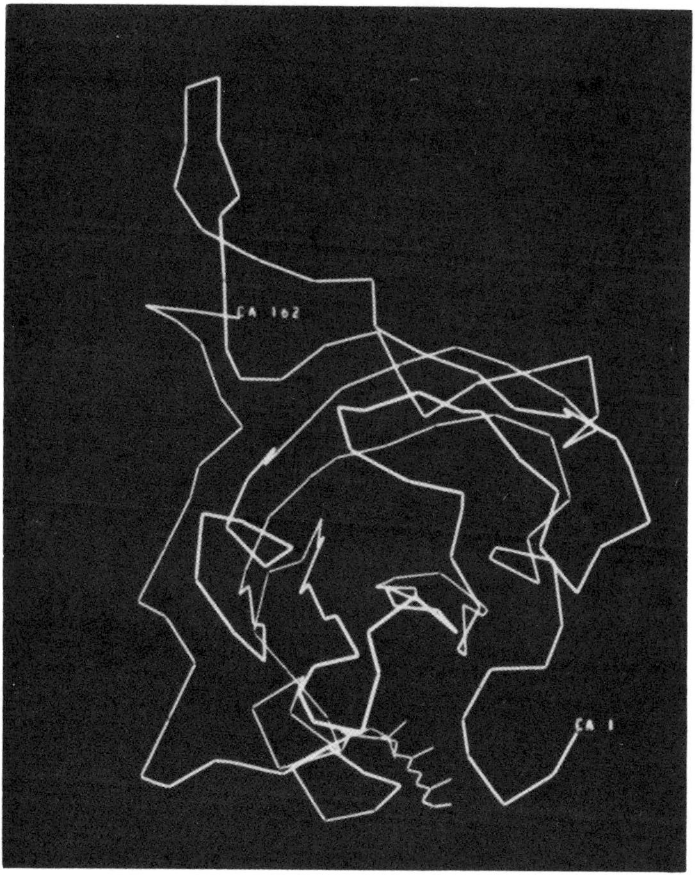

Figure 5 - Interaction of BLG with retinol (latter located at bottom center)

the trigonal crystals. Also in this case most of the residues pointing to the center of the barrel are hydrophobic with a couple of exceptions at the opening of the barrel.

We prepared the complex BLG-retinol by adding a very large excess of the vitamin dissolved in ethanol to an aqueous BLG solution at pH 7.5. The complex thus made yields yellowish fluorescent crystals which are isomorphous to those of the native protein. A complete data set of these crystals was collected at 2.5 Å resolution and a Fourier difference map using phases calculated from our final model of BLG was computed. Though the difference map is fairly noisy, there is one area which shows peaks significant enough to account for the presence of a retinol molecule. This area is shown in figure 5. Surprisingly enough, it is found not in the center of the barrel but in a pocket at the surface of the molecule which is limited by residues that are essentially all hydrophobic.

This unexpected result, raises many new questions because we have now found an interesting example of two molecules with limited sequence homology and extremely high structural homology that perform an analogous function in a significantly different way.

Acknowledgments. Our work on RBP was done in collaboration with R. Berni and S. Ottonello from the University of Parma. Our studies on BLG were done in collaboration with L. Sawyer from the University of Edinburgh and M. Bolognesi from the University of Pavia. We are grateful to Prof. M. Mammi for his continuous interest in this project.

References

1 - Goodman D.S. (1984) in The Retinoids (M.B. Sporn, A.B. Roberts and D.S. Goodman eds.) vol 2 p. 41-88. Academic Press London.

2 - Heller, J. (1975) J. Biol. Chem. 250 3613-3619.

3 - Chen, C.C. and Heller, J. (1977) J. Biol. Chem. 252 5216-5221.

4 - Peterson, P.A. (1971) J. Biol. Chem. 246 44-49.

5 - Berni, R., Ottonello, S. and Monaco, H.L. (1985) Anal. Biochem. 150 273-277.

6 - Rask, L., Anundi, H. and Peterson, P.A. (1979) FEBS Lett. 104 55-58.

7 - Newcomer, M., Jones, T.A., Aqvist, J., Sundelin, J., Eriksson, U., Rask, L., and Peterson, P.A. (1984) EMBO J. 3 1451-1454.

8 - Monaco, H.L., Zanotti, G., Ottonello, S. and Berni, R. (1984) J. Mol. Biol. 178 477-479.

9 - Hendrickson, W.A. (1985) in Methods in Enzymology (Wyckoff, H.W., Hirs, C.H.W. and Timasheff, S.N. eds.) Vol 115 B p. 252-270 Academic Press, London

10 - Futterman, S. and Heller, J. (1972) J. Biol. Chem. 247 5168-5172.

11 - Wishnia, A. and Pinder, T.W. Jr. (1966) Biochemistry 5 1534-1542.

12 - Wishnia, A. (1969) Biochemistry 8 5070-5075.

13 - Robillard, K.A. Jr. and Wishnia, A. (1972) Biochemistry 11 3835-3840.

14 - Palmer, A.H. (1934) J. Biol. Chem. 104 359-372.

15 - Hemley, R., Kohler, B.E. and Siviski, P. (1979) Biophys. J. 28 447-455.

16 - Fugate, R.D. and Song P.S. (1980) Biochim. Biophys. Acta 625 28-42

17 - Horwitz, J. and Heller, J. (1974) J. Biol. Chem. 249 4712-4719.

18 - Green, D.W., Aschaffenburg, R., Camerman, A., Coppola, J.C., Dunnill, P., Simmon, R.M., Komorowski, E.S., Sawyer, L., Turner, E.M. C. and Woods, K.F. (1979) J. Mol. Biol. 131 375-397.

19 - Sawyer, L., Papiz, M.Z., North, A.C. and Eliopoulous, E. (1985) Biochem. Soc. Trans. 13 265-266.

20 - Braunitzer, G., Chen, R., Schrank, B. and Stangl, A. (1973) Hoppe-Seyler's Z. Physiol. Chem. 354 867-878.

21 - Pervaiz, S. and Brew, K. (1985) Science 228 335-337.

STRUCTURAL BASES FOR THE RECOGNITION OF INHIBITORS BY SERINE PROTEINASES AND THEIR ZYMOGENS

Martino Bolognesi[+], Paolo Ascenzi[≠], Gino Amiconi[≠], Enea Menegatti[//] and Mario Guarneri[//]

[+]Department of Genetics and Microbiology, Section of Crystallography, University of Pavia, 27100 Pavia, Italy

[≠]C.N.R., Center for Molecular Biology, Department of Biochemical Sciences, University of Rome "La Sapienza", 00185 Rome, Italy

[//]Department of Pharmaceutical Sciences, University of Ferrara, 44100 Ferrara, Italy

The phenomenon of the recognition and binding of specific substances by specific binding sites is of fundamental importance in biology. From a static viewpoint, chemical recognition can easily be described in terms of complementarity; in other words, the interface between two interacting proteins, or between an enzyme and its substrate (or inhibitor) is stabilized by the close packing of atoms without internal holes, a maximum number of hydrogen bonds, pairing of oppositely charged groups and buried hydrophobic residues. A comparative study of the threedimensional structures of the complexes between serine (pro)enzymes and their inhibitors has been chosen as a type case for the illustration of these structural aspects of recognition. This kind of information, obtained by crystallographic studies (1), has been of primary importance in the understanding of the catalytic steps involved in the hydrolysis of peptide (or ester) bonds by serine proteinases. On the basis of kinetic and thermodynamic data, the sequence of reactions which are thought to describe the catalytic mechanism of these proteinases, shown in Fig. 1, includes: 1) enzyme:substrate productive association; 2) nucleophilic attack of the enzyme Ser 195 OG atom on the carbonyl carbon of the scissile peptide bond; 3) formation of an acyl-enzyme intermediate with departure of the amino portion of the cleaved peptide; 4) reversal of these steps (deacylation) with replacement of the second leaving product by an OH^- group provided by a water molecule (2).

A number of residues from the enzyme partecipate in these reaction steps; besides Ser 195, His 57 and Asp 102 play a key role in the so-called "charge-relay system", while others are more directly involved in proper orienta-

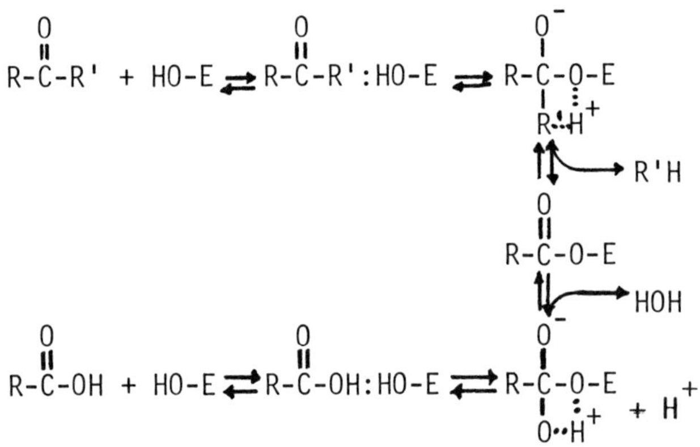

Fig. 1. Steps in the hydrolysis mechanism for peptide (or ester) substrates catalyzed by serine proteinases: R, R' substrate residue, E enzyme. Adapted from (2).

tion of the substrate during catalysis. Thus the main components of the catalytic and binding sites of serine proteinases comprise the "catalytic triad" (Asp 102, His 57, Ser 195), and the enzyme segments building up the S_1, S_2, ..., S_n sites (according to Schechter and Berger (3) the S_i site is the enzyme's binding site for residue P_i of the substrate or inhibitor). This general description is true not only for enzymes of the trypsin-chymotrypsin family, but also for the homologous microbial serine proteinases SGPA, SGPB (Streptomyces griseus protease-A and B) and α-lytic protease (from Myxobacter 495 (4)) as well as for subtilisin (5), taken as a prototype of an independent, but in a way evolutionary convergent family of proteolytic serine enzymes.

As mentioned above, the properties of natural and synthetic inhibitors which react with trypsin and related (pro)enzymes have provided information on the structural details and the binding modes of ligands to the active site of trypsin(ogen) (bovine). At least nine families of protein inhibitors of serine proteinases can be distinguished which obey to a "Standard Mechanism" of action, as proposed by Laskowski and Kato (6). Each member of a family displays extensive homology to other inhibitors of the same group, while no primary structure relation occurs between families. However on every inhibitor molecule one (or more) particular peptide bond (the "reactive site") can be distinguished, which specifically interacts with the active site of the target enzyme. The value of the so called k_{cat}/K_m ratio for the hydrolysis of the reactive site bond by the cognate enzyme is very high for these inhibitors, $10^4 - 10^6 M^{-1} s^{-1}$, although both k_{cat} and K_m are orders of magnitude lower than the corresponding values for normal substrates. Thus hydrolysis occurs very slowly and the enzyme:inhibitor system can be treated as a simple chemical equilibrium between the free E and I species and the E•I complex.

These properties lend themselves to, and indeed have

allowed, detailed crystallographic investigations of several enzyme:inhibitor complexes. These include: the structures of trypsin and anhydrotrypsin (bovine) with BPTI (bovine basic pancreatic trypsin inhibitor, Kunitz-type), of trypsinogen:BPTI in the presence and in the absence of the effector dipeptide Ile-Val (for a review see ref. 1), of trypsinogen:Arg^{15}BPTI:Val-Val(7) of trypsin: STI (soybean trypsin inhibitor, Kunitz-type (8)), of trypsinogen:PSTI (porcine pancreatic secretory trypsin inhibitor, Kazal-type (9)), and of kallikrein:BPTI (10). These structural data obviously parallel the crystallographic studies on free trypsin, trypsinogen and kallikrein (1, 11). In addition, if microbial serine proteinases are to be considered, the complexes of SGPB with the third domain of turkey ovomucoid inhibitor (12) as well as the subtilisin BPN':SSI adduct (Streptomyces subtilisin inhibitor (13)) should be mentioned.

These serine proteinase:protein inhibitor complexes are the only reliable models available at present which depict the binding of extended polypeptide segments to the active sites; they are therefore the starting elements for the study of the specific processes of enzyme:substrate recognition. It is worth mentioning that strong inhibitors of trypsin (such as BPTI and STI) are just substrates for a homologous trypsin 1 from a starfish (14). In order to stress this striking double role of the inhibitors the term "trapped substrate" has been proposed.

In what follows we shall focus mainly on the structures of PSTI and BPTI complexes with trypsin(ogen) and kallikrein, inspecting the binding modes of the inhibitors, the details of their specific interactions with the proteinases, and the possible roles of structural elements in determining the inhibitory (other than substrate) function. In addition, on the basis of the binding principles inferred from observation of the complexes with natural polypeptide inhibitors, a short discussion will follow on the interpretation (or else on the prediction), on structural grounds, of the binding behaviour of low molecular weight synthetic inhibitors of serine proteinases.

BINDING OF PSTI TO TRYPSINOGEN

The three-dimensional structure of the complex between PSTI and trypsinogen has been solved and refined to 1.8 Å resolution, with a final crystallographic R factor of 19.6% (9). Fig. 2 gives an overall view of the polypeptide backbone of the complex which assumes the typical "mushroom shape", the inhibitor being the stalk. Because of the peculiar properties of the activation domain of trypsinogen, the reaction of the proenzyme with strong polypeptide inhibitors induces an extended conformational transition in the molecule which adopts a tertiary structure almost coincident with that of trypsin. This effect, which can be better understood on

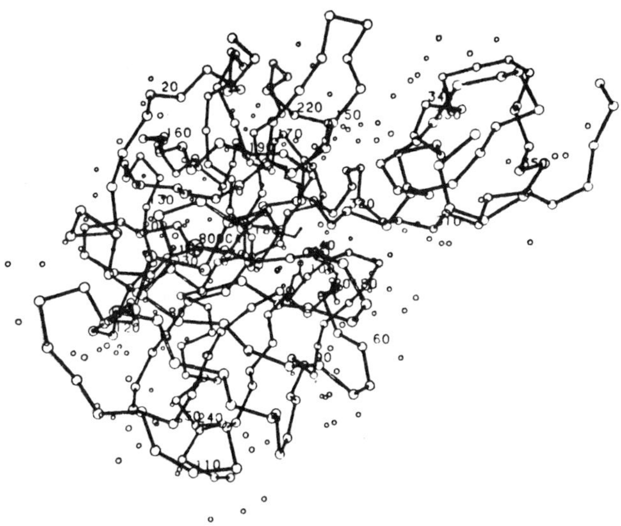

Fig. 2. Three-dimensional structure of the trypsinogen: PSTI adduct; ordered solvent water molecules have been drawn as isolated open circles.

the basis of the intramolecular activation mechanism of trypsin, has been extensively investigated from a functional viewpoint, by NMR and X-ray crystallography (1, 15, 16, 17). The zymogen:inhibitor complex, from a structural viewpoint, is thus a good model for the study of enzyme:inhibitor interactions, and in what follows we shall adopt the convention of referring to the "activated" structure of the proenzyme as "trypsin(ogen)".

The overall binding geometry of the PSTI complex resembles closely what has been observed in other proteinase:inhibitor adducts. PSTI is wedge shaped and binds to the active site pointing to the catalytic center. Tight compenetration of the two binding surfaces occurs such that the atomic structure at the proenzyme:inhibitor interface resembles that found in the inner core of globular proteins. The reactive site loop of PSTI (residues 10-21, sites P_9 through P_3') aligns in an antiparallel fashion with different backbone segments of the (pro)enzyme, giving rise to what might look as a rather extended stretch of β-structure (Fig. 3). As will be discussed later,

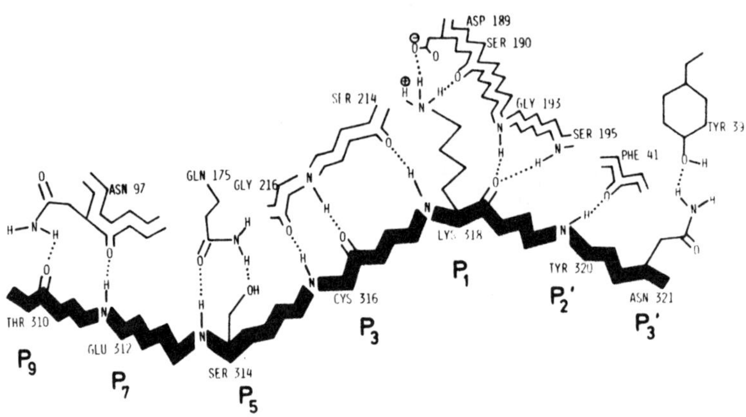

Fig. 3. Schematic representation of the (re)active site contact area in the trypsinogen:PSTI complex. The inhibitor polypeptide backbone is in black; residue numbering in PSTI can be distinguished by the digit 3 preceding the true sequence number of a given amino acid.

particularly interesting are the hydrogen bonds connecting the inhibitor to segments 39 - 41 and 214 - 216 of trypsin(ogen). The side chain of Lys 18 (PSTI P_1 residue) fits into the specificity pocket, and makes hydrogen bonds to internal W 414 and 416 and to Ser 190; the positive charge of the ε-ammonium group is counterbalanced by the carboxylate of Asp 189 of the (pro)enzyme. Another relevant inhibitor:trypsin(ogen) interaction connects (by means of a bifurcated H bond) the carbonyl oxygen of the scissile peptide bond (18 - 19 in PSTI) to the "oxyanion hole", i.e. the peptide nitrogens of residues Gly 193 and Ser 195. Thus a total of 11 hydrogen bonds connects donors/acceptors of the two molecules, most of them are linear and occur between main chain atoms. To these the specific interactions at the S_1 site must be added.

In the crystal structure seven tightly bound water molecules mediate the trypsinogen:inhibitor interaction, five of which in the reactive site loop. Of these W 403 is at the P_3 site where it makes hydrogen bonds bridging residues Cys 16 (PSTI) with Glu 192 and Gly 216 of the (pro)enzyme. At site P_5 W 711 connects Ser 14 (PSTI) with Lys 224 (trypsinogen), while at site P_3' W 604 links Asn 21 (PSTI) to Tyr 39 (proenzyme). Remarkably the solvent molecules 414, 416 and 403 fall in exactly equivalent positions in the various trypsin: or trypsinogen:BPTI complexes studied (1) despite the structural differences between the two inhibitors.

A second contribution to the energetics of the binding process comes from the van der Waals intermolecular contacts. The structural complementarity between the two interacting surfaces is high: 16 residues from PSTI participate in the 138 intramolecular contacts (shorter than 4.0 Å) observed in the complex. For purposes of comparison with other inhibitors, therefore, the following six major areas of enzyme:inhibitor recognition can be distinguished (Fig. 3): 1) the S_9 - S_5 contact region;

2) the S_3-S_1 backbone interactions; 3) the specificity pocket; 4) interactions at the S_1'-S_3' sites ; van der Waals interactions; 6) seven (possibly more) bridging water molecules.

The reactive site loop of Kazal inhibitors is limited by two disulphide bridges (16-35 and 24-56) which keep this exposed segment of polypeptide structure close to the central alpha helix. Antiparallel to the inhibitor reactive site backbone runs a short segment of polypeptide structure (res. 30-35) which precedes the N-terminal end of the helix (Fig. 4). Residue Asn 33, from this region,is highly conserved in all the Kazal inhibitors primary structures known; its side chain exploits fully its hydrogen bonding capabilities (3 H bonds) and acts as a linker/spacer between the reactive site and the inner (rigid) core of the inhibitor. In the complex the 30-35 loop falls rather near to the enzyme surface and has been indicated as a possible secondary contact area. Indeed in the PSTI: trypsinogen complex one contact of 3.8 Å is observed between the side chain of residues Ser 32 of the inhibitor and Gln 192 of the (pro)enzyme, while substitutions at residue 32 (PSTI) with Gly or with bulky residues have been shown to affect the binding (18). Weak interactions at the secondary contact area have been observed in the homologous complex between SGPB and the third domain of turkey ovomucoid inhibitor (12), and to a higher extent in the subtilisin BPN':SSI complex (13). A topologically equivalent secondary contact area can be identified also in BPTI (1, 10).

Once the high resolution structure of an enzyme:inhibitor complex is available it is of a certain interest to examine it in terms of the sequences of closely homologous molecules and of their inhibition constants. The bovine pancreatic secretory trypsin inhibitor (b.PSTI) shows substitutions at 12 positions, as compared to the porcine inhibitor. Three of these occur at the interface with the enzyme at sites P_8, P_5 and P_1, where Ser 11, Ser

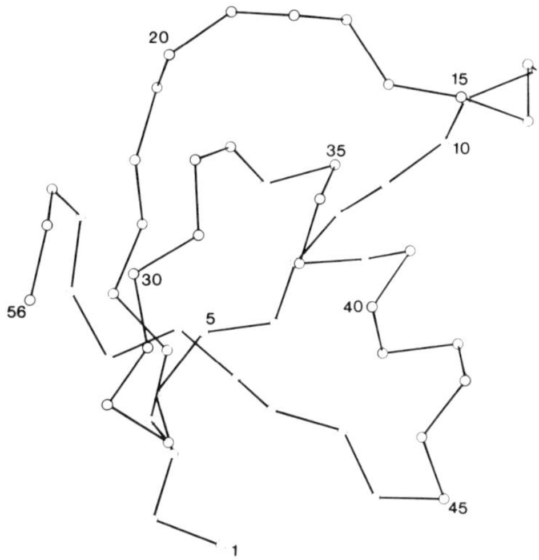

Fig. 4. C-alpha schematic drawing of the polypeptide backbone of the Kazal inhibitor. In this orientation the reactive site loop is in the upper part of the figure.

Fig. 5. C-alpha schematic drawing of the polypeptide backbone of the trypsin:BPTI complex.

14 and Lys 18 are replaced respectively by Asn, Asn and Arg. Inspection of the three-dimensional model of the complex indicates how these changes are likely to affect the enzyme:inhibitor recognition pattern. In particular at the S_1 site the guanidino group of Arg 18 (b.PSTI) removes the buried water molecule W 414 and makes H bonds to O 219, OG 190, W 416 and to the carboxylate group of Asp 189.

Residues P_5 in b.PSTI is larger than in the porcine inhibitor, and can reach the NZ atom of trypsin Lys 224, introducing thus little changes in the enzyme:inhibitor binding pattern. The other substitutions observed in b.PSTI occur in areas far from the contact region and exposed to solvent; thus they are not likely to act as modulators of the affinity for trypsin(ogen). It should be pointed out however that extensive substitutions are observed in the PSTI N-terminus (including apolar residues and one Gly). Even though it is possible that this part of the molecule adopts a novel conformation in b.PSTI, as compared to PSTI, De Marco et al. (19) have shown that removal of N-terminal fragments up to residue 8 does not affect the inhibitory properties of PSTI. This obser-

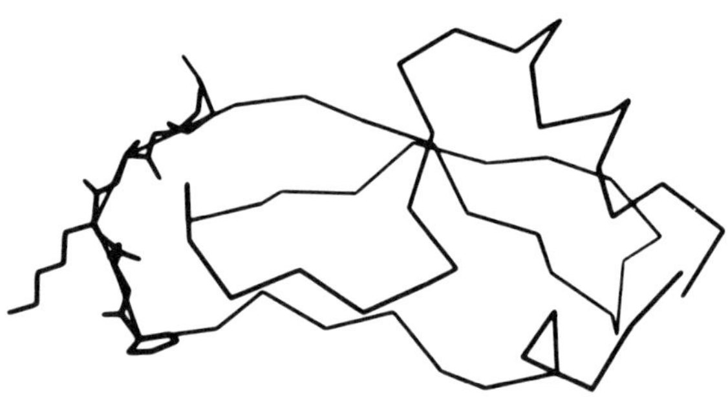

Fig. 6. The C-alpha skeleton of basic pancreatic trypsin inhibitor. In the reactive site loop the full structure of the inhibitor has been added. The side chain of the reactive site Lys 15 is clearly visible.

vation is in accordance with the absence of (pro)enzyme: inhibitor contacts in this region (see Fig.2).

BINDING OF BPTI TO TRYPSIN AND KALLIKREIN

The high affinity of BPTI for both trypsin and kallikrein has allowed the growth and analysis of stable inhibitor adduct crystals (1, 10). Owing to the high homology of the two serine proteinases the complexes share several common structural features which are summarized in this section. Fig. 5 shows that the complexes can be described again as "mushroom shaped". The BPTI tertiary structure is illustrated in Fig. 6, with the reactive site Lys 15 residue highlighted at the top. Comparison of Fig. 4 and 6 indicates that PSTI and BPTI are structurally unrelated and that their reactive site loops have quite different curvatures. The distances between P_4 and P_3' residues, as measured from the respective C-alpha atoms, are 11.6 Å in BPTI and 20.0 Å in PSTI. This fact is in agreement with the observation of a reduced number of intermolecular H bonds in the (re)active site area of the trypsin: BPTI complex. Nevertheless, if the structures of the polypeptide backbones are compared in the strict neighborhood of the reactive bonds, remarkable coincidence of the ψ, φ conformational angles is observed. This is true also for the free inhibitors, indicating that very little readjustement is required for the inhibitors to adapt to the enzymes' active sites (predetermined structural complementarity).

Fig. 7 compares the enzyme:inhibitor polar contacts at the primary binding sites for the PSTI complex and for trypsin:BPTI. The pattern shown for trypsin:BPTI can be taken as representative of the kallikrein:BPTI complex as well, although, because of substitutions, different interactions occur at site P_2 (hydrophobic in kallikrein) and at P_4'.

The reactive site loop of BPTI reverses its direction

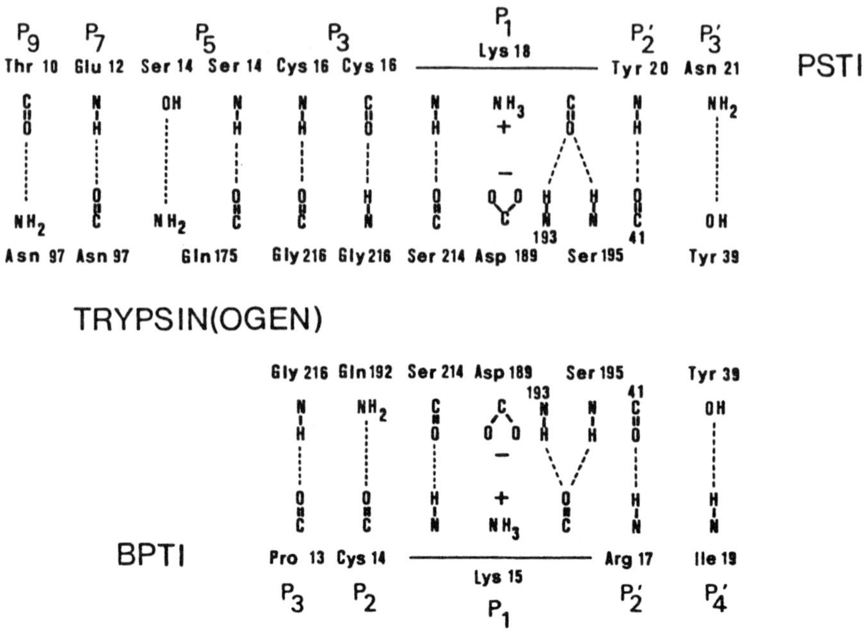

Fig. 7. Comparison of hydrogen bond interactions at the primary contact areas of PSTI: and BPTI:trypsin(ogen) adducts.

rather abruptly; as a result sites $P_9 - P_5$ (residues 7 through 11) cannot reach the enzyme and no recognition occurs in this area, as opposed to what has been found for Kazal inhibitors and in the SSI:subtilisin complex. Thus, five out of the six major areas of enzyme:inhibitor recognition listed in the previous section are present: 1) the $S_1 - S_3$ contact region; 2) the specificity pocket; 3) interactions at sites $S_1' - S_3'$; 4) van der Waals interactions; 5) bridging water molecules. In trypsin:BPTI 121 atomic contacts are found, while in the kallikrein complex 138 van der Waals interactions (in both cases shorter that 4.0 Å) are provided by inhibitor residues. Solvent water molecules W 414 and W 416 are present in structural equivalent positions at the S_1 sites of both enzymes, and mediate the polar and electrostatic interactions of the ammonium group of P_1 Lys 15 residue. Aminoacid substitutions in the specificity pocket of trypsin, compared to kallikrein (Gly 226→Ser 226; Ser 190→Thr 190) and minor backbone readjustements have some influence on the conformation of the P_1 residue side chain in the two enzymes, possibly indicating one structural reason for the preference of kallikrein for arginine at the P_1 site of substrates and inhibitors.

BINDING OF SYNTHETIC INHIBITORS TO SERINE PROTEINASES

The availability of refined crystallographic structures of serine proteinases allows the study of related enzymes by means of the so called "homology modeling" technique (20). Within certain limits it is possible to build regions of serine proteinases of unknown three dimensional structures, on the basis of their primary structures and taking advantage of computer graphics methods and of the crystallographic protein structures data base. Once a homologous model is obtained, functional properties towards ligands can be analyzed in conjunction with the computer-generated structure most reliably in those areas of the enzyme where limited substitutions, deletions or insertions have occurred compared to the crystallographic starting model(s). According to these lines we have examined the binding modes of two synthetic inhibitors of serine proteinases, which have potential use as drugs in the treatment of diseases related to uncontrolled activity of plasmatic or pancreatic serine proteinases (for example disseminated intravascular coagulation and acute pancreatitis (21)).

The first compound considered (ethyl-p-(6-guanidino-hexanoyloxy) benzoate methane sulphonate, FOY®) is structurally reminiscent of lysine and arginine side chains, and shows the highest inhibitory effect towards cationic serine proteinases. The competitive inhibitory effect of the drug

Fig. 8a. The predicted binding mode of the compound FOY® to trypsin.

was measured in terms of its inhibitory constants for trypsin, (porcine pancreatic) and (human urinary) kallikrein, (human) urokinase, (human) Lys^{77} plasmin, (bovine) factor Xa and α-thrombin. The measurements showed that FOY® inhibition pattern for the enzyme considered is rather anomalous: in fact the drug does not show higher affinities for the enzymes whose primary specificity is known to be for arginyl residues. In fact, the observed values of the inhibitory constants (21) indicate that the affinity of the drug is lowest for the two kallikreins.

Model building experiments conducted indicated that FOY® can bind to the active site of serine proteinases in a manner fully compatible with the stereochemical rules obeyed by other inhibitors whose crystallographic structure is known (see above). In particular the following three points characterize the enzyme:inhibitor interaction: 1) insertion of the aliphatic tail of FOY® in the S_1 site of the proteinases, with formation of typical H bonds and charge interaction with the carboxylate group of Asp 189; 2) the carbonyl group of the "scissile" ester bond of the drug aligns properly and accommodates its oxygen atom in the "oxyanion hole" of the enzyme, where it makes hydrogen bonds to the amide nitrogens of residues Gly 193 and Ser 195; 3) minor interactions, if any, occur at the ethyl benzoate moiety of the inhibitor, which points to the solvent region. These interactions involve aminoacid residues which are highly conserved in all the proteinases considered, and should not be the source of the unusual specificity of FOY®. Inspection of the models shows that the aliphatic chain supporting the guanidino group is one carbon atom longer in FOY® than in arginine, and occupies to a different extent the specificity pocket of the enzymes. This observation stresses the role played by the residues lining up along the walls of the S_1 site of the different proteinases. Analysis of the respective primary structures (22) and model building experiments conducted in this area indicate that the structural effects of a substitution at residue 226 can be decisive for the enzyme interaction with a guanidino group supported by an aliphatic chain longer than that of arginine. This position is occupied by a glycine in trypsin, α-thrombin, factor Xa, plasmin and urokinase, while it is a seri-

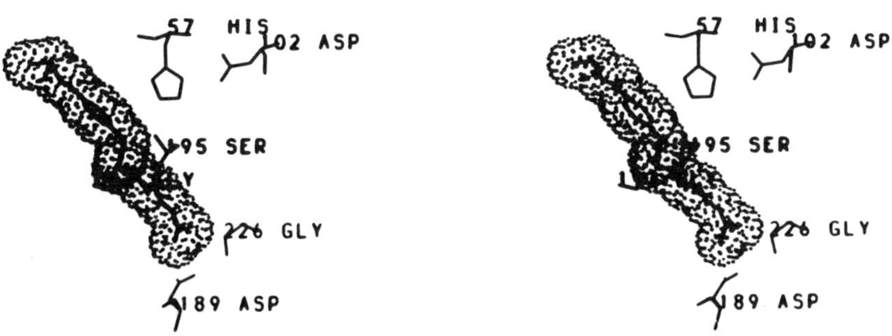

Fig. 8b. Details of the FOY®: trypsin interactions at the enzyme specificity pocket.

ne in kallikrein. Fig. 8a shows a schematic model of the C-alpha backbone of trypsin with bound inhibitor; in Fig. 8b, and c, respectively, are shown the catalytic triad and the relevant S_1 residues for the active sites of a trypsin-like (Gly 226) proteinase and of kallikrein (Ser 226).

A comparable study was conducted on the inhibitor tetra-p-aminophenoxy-neo-pentane (TAPP), which was synthesised (23) keeping in mind the well known inhibitory effect of benzamidine towards serine proteinases (24). TAPP shows several relevant biological effects, among which the inhibition of *in vitro* fibrinogen-to-fibrin conversion (see (21) and references cited therein). TAPP binding to serine proteinases shows peculiar properties which can be related to the "multihead" nature of this inhibitor. In fact, besides showing different affinities for different enzymes, in the case of kallikrein the pH profile of the inhibitory constants indicates that inhibitor association depends from two equivalent TAPP-linked acid groups.

As outlined above for the FOY® inhibitor, computer generated models and available crystallographic structures of the serine enzymes were employed in a systematic analysis of TAPP binding modes. The main enzyme:inhibitor interaction occurs at the specificity site, and was modeled according to the coordinates of both trypsin:benzamidine and kallikrein:benzamidine crystallographic analyses (11, 25). As shown in Fig. 9 once a benzamidine moiety is bound to the S_1 site the additional three fall close to the enzyme surface and have a certain conformational freedom to explore it "looking for additional interactions". A systematic search of a surface charged group potentially useful for binding of a second benzamidine substituent was therefore conducted on a graphic device in conjunction with the primary structures of the enzymes, and allowing for proper torsional degrees of freedom in the TAPP molecule. Such search showed that, in accordance with the thermodynamic data, no negatively charged group can be reached

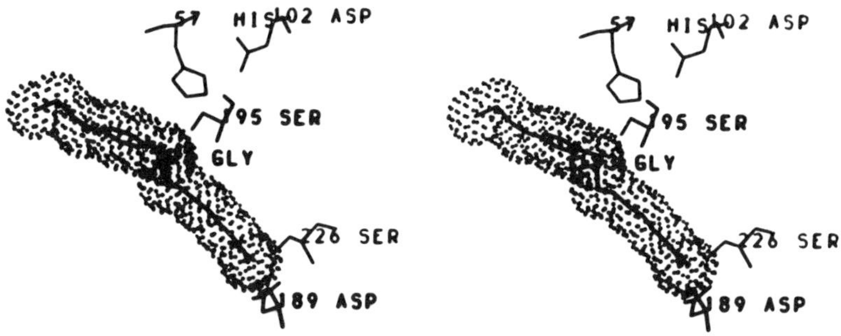

Fig. 8c. FOY® : kallikrein interactions at the specificity site (compare with Fig. 8b).

by the inhibitor substituents in trypsin as secondary interaction. Notwithstanding the availability of the high resolution crystal structure of kallikrein it was not possible to find a unique solution for the search of the secondary binding group. Both Asp 148 as well as Asp 174 can

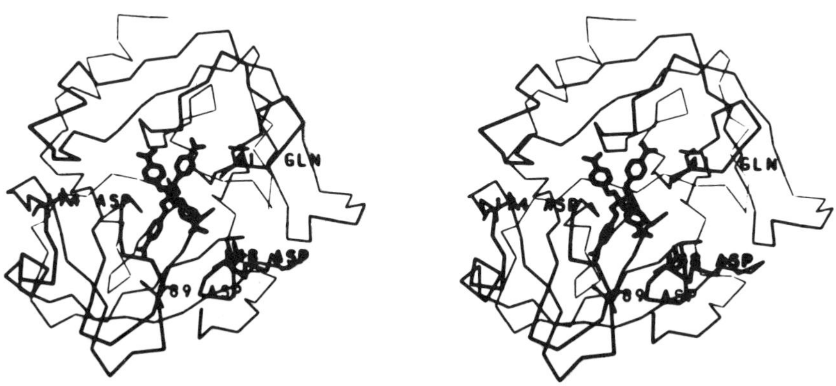

Fig. 9. Binding mode of TAPP to kallikrein.

be alternatively reached by a TAPP benzamidine moiety, with acceptable conformational readjustments (see Fig. 9). The higher inhibitory effect of TAPP on kallikrein, as compared to trypsin, may also be explained on the basis of pH independent additional interactions between the inhibitor substituents and the enzyme in the neighborhood of the central site.

CONCLUSIONS

Most of the specificity in a protein-protein interaction must reside in the structures of the proteins themselves; therefore, the understanding of the recognition-transformation process in proteinases at a molecular level requires an accurate knowledge of the spatial distribution of the chemical groups and their geometrical disposition at level of the interacting surfaces.

The reported experimental evidence indicates that the residues at the interface between serine (pro)enzyme(s) and their inhibitor(s) are well interdigitated and that the positions of the corresponding knobs and holes match as well. In other words, from a structural point of view, recognition and specificity appear to be simply an expression of how well chemical groups and forces match at the interacting surface. Such structural complementarity is mostly predetermined; in fact, in the formation of the adducts, only small readjustements of the conformation of the free enzyme and of the inhibitor are required.

REFERENCES

1. Huber, R., Bode, W. (1978) Acc. Chem. Res. 11, 114-121.
2. Kraut, J. (1977) Annu. Rev. Biochem. 46, 331-358.
3. Schechter, I., Berger, A. (1967) Biochem. Biophys. Res. Commun. 27, 157-162.
4. James, M.N.G., Sielecki, A.R., Brayer, G.D., Delbaere,

L.T.J., Bauer, C.A. (1980) J. Mol. Biol. 144, 43-88.
5. Robertus, J.D., Alden, E.A., Birktoft, J.J., Kraut, J., Powers, J.C., Wilcox, P.E. (1972) Biochemistry 11, 4293-4303.
6. Laskowski, M. Jr., Kato, I. (1980) Annu. Rev. Biochem. 49, 593-626.
7. Bode, W., Walter, J., Huber, R., Wenzel, H.R., Tschesche, N. (1984) Eur. J. Biochem. 144, 185-190.
8. Sweet, R.M., Wright, H.T., Janin, J., Chothia, C.N., Blow, D.M. (1974) Biochemistry 13, 4212-4228.
9. Bolognesi, M., Gatti, G., Menegatti, E., Guarneri, M., Marquart, M., Papamokos, E., Huber, R. (1982) J. Mol. Biol. 162, 839-868.
10. Chen, Z., Bode, W. (1983) J. Mol. Biol., 164, 283-311.
11. Bode, W., Chen, Z., Bartels, K., Kutzbach, C., Schmidt-Kastner, G., Bartunik, H. (1983) J. Mol. Biol., 164, 237-282.
12. Read, R.J., Fujinaga, M., Sielecki, A.R., James, M.N.G. (1983) Biochemistry 22, 4420-4433.
13. Hirono, S., Akagawa, H., Mitsui, Y., Iitaka, Y. (1984) J. Mol. Biol. 178, 389-413.
14. Estell, D.A., Laskowski, M. Jr. (1980) Biochemistry 19 124-131.
15. Ascenzi, P., Amiconi, G., Bolognesi, M., Menegatti, E., Guarneri, M. (1985) Biochim. Biophys. Acta 832, 378-382.
16. Menegatti, E., Guarneri, M., Bolognesi, M., Ascenzi, P., Amiconi, G. (1985) Biochim. Biophys. Acta 832, 1-6.
17. Perkins, S.J., Wuthrich, K. (1980) J. Mol. Biol. 138, 43-64.
18. Bogard, W.C. Jr., Kato, I., Laskowski, M. Jr. (1980) J. Biol. Chem. 225, 6569-6574.
19. De Marco, A., Menegatti, E., Guarneri, M. (1982) Biochemistry 21, 222-229.
20. Jones, T.A. (1985) Methods in Enzymology 115, 157-170.
21. Menegatti, E., Guarneri, M., Bolognesi, M., Ascenzi, P., Amiconi, G. (1987), accompanying paper in this volume.

22. Menegatti, E., Bolognesi, M., Scalia, S., Bortolotti, F., Guarneri, M., Ascenzi, P. (1986) J. Pharm. Sci. in the press.
23. Ferroni, R., Menegatti, E., Guarneri, M., Taddeo, U., Bolognesi, M., Ascenzi, P., Amiconi, G. (1984) Il Farmaco 39, 901-909.
24. Mares-Guia, M., Shaw, E. (1965) J. Biol. Chem. 240, 1579-1585.
25. Bode, W., Schwager, P. (1975) J. Mol. Biol. 98, 639-717.

ROLE OF THE PRIMARY SPECIFICITY SUBSITE ON THE INTERACTION BETWEEN SERINE PROTEINASES AND LOW MOLECULAR WEIGHT SUBSTRATES AND INHIBITORS

Enea Menegatti[+], Mario Guarneri[+], Martino Bolognesi[§], Paolo Ascenzi[*] and Gino Amiconi[*]

[+]Department of Pharmaceutical Sciences, University of Ferrara, 44100 Ferrara, Italy
[§]Department of Genetics and Microbiology, Section of Crystallography, University of Pavia, 27100 Pavia, Italy
[*]C.N.R., Center for Molecular Biology, Department of Biochemical Sciences, University of Rome 'La Sapienza', 00185 Rome, Italy

The concept of stereochemical complementarity, which is fundamental to our understanding of the replication of genetic material, has been invoked to explain molecular specificity in enzymes, where the structural match is realized either at the initial stage of binding or later during the formation of a transition complex. Such models are inadequate to expound some aspects of the binding, recognition and selectivity processes in which dynamical factors cannot be disregarded. However, fitting requirements are sufficient to fully explain small specificity variations and to partially understand the role played by local parts of a molecular structure in the overall functional process of interest. These ideas are nicely illustrated in the interaction between the primary specificity subsite of serine proteinases and low molecular weight substrates and inhibitors.

KINETICS FOR THE HYDROLYSIS OF N-α-CARBOBENZOXY-L-ARGININE p-NITROPHENYL ESTER AND N-α-CARBOBENZOXY-L-LYSINE p-NITROPHENYL ESTER BY PORCINE PANCREATIC β-KALLIKREIN-B, BOVINE α-THROMBIN AND BOVINE β-TRYPSIN

Extensive kinetic and thermodynamic studies have shown that the arginine residue is preferred to the lysine one at the primary specificity subsite (i.e., S_1) of porcine (p.) pancreatic (p.) β-kallikrein-B (1) and of bovine (b.) α-thrombin (2). On the other hand, bovine (b.) β-trypsin undistinguishably catalyzes the hydrolysis of synthetic substrates containing an arginine or a lysine residue at the P_1 position (1,2). In order to clarify the role played by the different amino acid residues at the S_1 subsite in the catalytic behaviour of serine proteinases acting on cationic substrates, the kinetics for the hydrolysis of N-α-carbobenzoxy-L-arginine p-nitrophenyl ester (ZArgONp) and N-α-carbobenzoxy-L-lysine p-nitrophenyl ester (ZLys-ONp) (see Figure 1) catalyzed by p. p.β-kallikrein-B, b. α-thrombin and b. β-trypsin (1,2) have been analyzed in parallel, considering at the same time the threedimensional structures of the various enzymes (2-7).

Over the whole pH range explored (pH 2.0 - 8.0), at I=0.1 M and 21°C, the hydrolysis of ZArgONp and ZLysONp catalyzed by p. p. β-kallikrein-B, b. α-thrombin and b. β-trypsin conforms to simple Michaelis-Menten kinetics, and pre-steady-state and steady-state parameters may be consistently fitted to the minimum three-step mechanism involving the acyl enzyme E·P (Scheme I):

$$E + S \underset{}{\overset{K_s}{\rightleftharpoons}} E\cdot S \xrightarrow{k_{+2}} \underset{\displaystyle P_1}{\underset{+}{E\cdot P}} \xrightarrow{k_{+3}} E + P_2$$

where E is the enzyme, S is the substrate, E·S is the rapidly formed enzyme · substrate adduct, and P_1 and

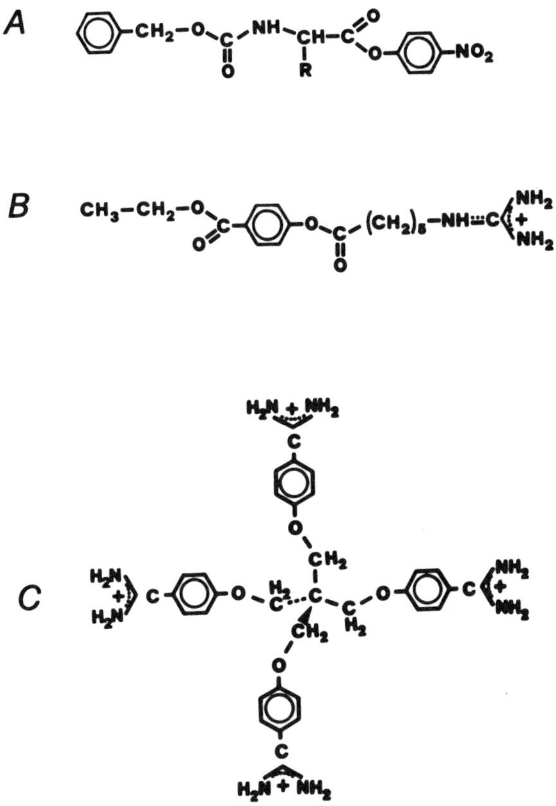

Figure 1. Chemical structures of ZArgONp (A, R = $(CH_2)_3$-NH⋯$(NH_2)C{:}NH_2^+$), ZLysONp (A, R = $(CH_2)_4$-NH_3^+), FOY® (B) and TAPP (C).

P_2 are the hydrolysis products.

All the enzyme/substrate systems considered imply two catalytically relevant residues, possibly His 57 and Asp 189 (2-8), acting in the neutral and acidic regions, respectively. However, while the imidazole group of His 57 does not undergo any pK_a shift upon the formation and disappearance of the E·S and E·P complexes, protons are released or taken up from p. p. β-kallikrein-B, b. α-thrombin and b. β-trypsin below pH 5.5 as a consequence of the pK_a perturbation of the Asp 189 side chain on substrate binding and product dissociation (see Table 1).

As far as the pH profiles of the kinetic parameters for the enzymatic hydrolysis of ZArgONp and ZLysONp are concerned, significant differences among the three proteinases are observed in the acidic pH region. Such differences are especially evident in the proton dependence of the deacylation step (k_{+3}) (Table 1). The sequence of events associated with the k_{+3} step appears to be similar for b. α-thrombin and b. β-trypsin catalyzed hydrolysis of ZArgONp and ZLysONp as well as for the p. p. β-kallikrein-B catalyzed hydrolysys of ZLysONp. Thus, from the functional viewpoint, the restoration of free enzymes is preceded or accompanied by the disappearance of the Asp 189 residue, possibly its pK_a being forced below pH 2.0 during the dissociation of the E·P adduct, or rather because it does not contribute at all to the k_{+3} step. On the other hand, with ZArgONp, below pH 5.5, the restoration of free p. p. β-kallikrein-B is accompanied by the protonation of the Asp 189 side chain. The findings may be taken as indicative of structural changes in the S_1 subsite that depend on the nature of the substrate and are associated with the k_{+3} step. This explanation is based on the assuption that the same titrable group affects the pH profiles of k_{+2}/K_S ($=k_{cat}/K_m$), k_{+2} and k_{+3} which reflect ionization(s) in the free enzyme (E) and in

Table 1. pK_a and Mid-point Values Fitting Kinetics for the Hydrolysis of ZArgONp and ZLysONp (T = 21°C).

Kinetic Parameters[a]	**P. p. β-kallikrein-B**		**B. α-thrombin**		**B. β-trypsin**	
	ZArgONp	**ZLysONp**	**ZArgONp**	**ZLysONp**	**ZArgONp**	**ZLysONp**
k_{cat} (pK_a)	4.2	4.4	4.5	4.4	4.6	4.6
	7.0	6.9	6.8	6.9	7.1	7.1
k_{+2} (pK_a)	3.4	2.9	3.8	3.6	3.73	3.73
	7.05	6.9	6.8	6.9	7.05	7.05
k_{+3} (pK_a)	4.4	—	—	—	—	—
	7.0	6.7	6.6	6.6	6.95	6.95
K_s (mid-point)	3.8	3.6	4.15	4.0	4.2	4.2
k_{+2}/K_s ($=k_{cat}/K_m$) (pK_a)	4.2	4.35	4.5	4.4	4.5	4.5
	7.05	6.9	6.8	6.9	7.1	7.1

[a]For all the enzyme/substrate systems considered, values of K_m are pH independent (1,2).

the E·S and E·P adducts, agrees with the view that conformational transitions occur at the active site during catalysis (1,2,9).

Although transient structural changes cannot be interpreted directly on the basis of the available molecular models (2-7), the differences in the catalytic behaviour and primary specificity of the three proteinases towards ZArgONp and ZLysONp are likely to be related to different enzyme/substrate interactions occurring at the S_1 subsite. Among the amino acid side chains involved in the recognition and binding of the substrate P_1 residue, changes at positions 190, 221 and 226, as well as in the 217-219 segment, may play a relevant role (2). Thus, Thr 190 in p. p. β-kallikrein-B and Ser 190 in b. β-trypsin are hydrogen bonded to the Lys 15 NE atom of the bovine basic pancreatic trypsin inhibitor. The presence of a smaller residue (i.e., Ala) at position 190 in b. α-thrombin allows an easier accomodation of the bulky arginine side chain at P_1. In addition, Asp 221 in b. α-thrombin (Gly and Ala in p. p. β-kallikrein-B and b. β-trypsin, respectively) may form hydrogen bond(s) with the substrate arginine guanidino group. Similarly, Ser 226 in p. p. β-kallikrein-B (Gly in b. α-thrombin and b. β-trypsin) is likely to affect the binding of the lysine ammonium group of the substrate. Moreover, cis-Pro 219 in p. p. β-kallikrein-B (Gly in b. α-thrombin and b. β-trypsin) alters the conformation of the neighbouring residues, by widening the S_1 subsite, thus allowing the preferential binding of the arginine side chain.

Although possible additional subtle structural sources of enzyme/substrate specificity cannot be ruled out, the data reported here indicate that different specificity and/or kinetic behaviour(s) of the serine proteinases considered can be related to different single amino acid replacements at their S_1 subsite.

THERMODYNAMICS OF FOY® INHIBITION OF SERINE PROTEINASES

FOY® is a non-antigenic synthetic inhibitor of plasmatic and pancreatic serine proteinases, that is used therapeutically in the treatment of pancreatitis, of disseminated intravascular coagulation and as a regional anticoagulant for hemodialysis (10,11). The use of FOY® as a drug stimulated our interest in the molecular basis underlying the formation of the enzyme: inhibitor binary adducts (11). To this end, the inhibition constant (K_i) for the binding of FOY® to a number of plasmatic, urinary and pancreatic serine proteinases all acting on cationic substrates, namely bovine (b.) factor Xa, b. α-thrombin, human (h.) Lys^{77}-plasmin, human (h.) urinary (u.) kallikrein, human (h.) urokinase, p. p. β-kallikrein-B3 and b. β-trypsin, has been determined (11).

For all the enzyme/inhibitor systems considered the FOY® inhibition pattern is strictly competitive. Moreover, titrations conform to simple equilibria, and the values of K_i are independent of the proteinase concentration.

Table 2 shows the K_i values, at pH 6.8 and 20°C, for the binding of FOY® to b. factor Xa, b. α-thrombin, h. Lys^{77}-plasmin, h. u. kallikrein, h. urokinase, p. p. β-kallikrein-B and b. β-trypsin (11). The data listed in Table 2 indicate that the affinity of FOY® for the serine proteinases examined does not follow the primary specificity of each enzyme for the ε-amino and/or for the δ-guanidino group(s) of lysyl or arginyl residues, respectively, that are at the P_1 position of substrates and/or inhibitors. In fact, on the basis of the primary specificity of the serine proteinases considered, FOY®, whose positively charged side chain is one carbon atom longer than the arginyl residue (see Figure 1), could be expected to show the highest affinity for h. u. kallikrein, p. p. β-kallikrein-B,

b. factor Xa and b. α-thrombin, which hydrolyze preferably arginyl side chains (1,2,12-15), an intermediate affinity for b. β-trypsin, which does not discriminate between arginyl and lysyl residues (2), and the lowest affinity for h. urokinase and h. Lys^{77}-plasmin, both of which catalyze preferably the cleavage of lysyl side chains (2,13,14).

In contrast, the experimental K_i values shown in Table 2 indicate that the affinity of FOY® for the various serine proteinases can be arranged as follows: h. u. kallikrein ~ p. p. β-kallikrein-B > b. β-trypsin ~ b. factor Xa ~ h. Lys^{77}-plasmin ~ h. urokinase ~ b. α-thrombin. These findings can be accounted for on the basis of the different interactions between the enzymes and FOY® occurring at the S_1 subsite (11) disclosed by a computer graphics model of the binding mode of FOY® to the proteinases. Comparison of the arginine side chain with the alyphatic tail of FOY®, supporting the ε-guanidino group, shows that the latter is one carbon atom longer (see Figure 1). This stereochemical constraint has the double effect of pushing the ε-guanidino group somewhat deeper into the primary specificity pocket and of requiring more room for the alyphatic chain along the pocket itself. It is worth nothing that residue 226, situated along the wall of the S_1 subsite, is Gly in b. factor Xa, b. α-thrombin, h. Lys^{77}-plasmin, h. urokinase and b. β-trypsin (3,4,16, 17). On the other hand, residue 226 is Ser in p. p. β-kallikrein-B and presumably also in the homologous h. u. kallikrein (18).

Model building observations indicated that the structural effect of a substitution at position 226 can be decisive in the enzyme interaction with the ε-guanidino alyphatic chain of FOY®. In conclusion, although other subtle structural sources of enzyme:FOY® complex (de)stabilization cannot of course be ruled out, the data reported here indicate that the optimal

Table 2. Values of the Inhibition Constant (K_i) for the Binding of FOY® to Serine Proteinases (pH 6.8; T = 20°C).

Proteinase	K_i (M)
H. u. kallikrein	2.6×10^{-3}
P. p. β-kallikrein-B	5.0×10^{-4}
B. β-trypsin	2.6×10^{-6}
B. factor Xa	2.2×10^{-6}
H. Lys^{77}-plasmin	1.5×10^{-6}
H. urokinase	1.3×10^{-6}
B. α-thrombin	8.0×10^{-7}

fit of FOY® at the S_1 subsite of the serine proteinases considered depends principally on specific amino acid distribution provided by the single enzymes at their specificity center.

THERMODYNAMICS OF TETRA-p-AMIDINOPHENOXY-*neo*-PENTANE INHIBITION OF SERINE PROTEINASES

The aromatic tetrabenzamidine derivative, tetra-p-amidinophenoxy-*neo*-pentane (TAPP), is a synthetic inhibitor of serine proteinases (19-22) which inhibits: (i) *in vivo*, the growth of sarcoma 180 implanted in hybrid mice (19); and (ii) *in vitro*, the blood coagulation induced by the "cancer coagulation factor" produced by the Walker carcinoma in Wistar rats (19), the fibrinogen-to-fibrin conversion (20), and the hydrolysis of esters and anilides of L-amino acids catalyzed by serine proteinases acting on cationic and non-cationic substrates (20-22). The stronger inhibitor effect of TAPP as compared to that of benzamidine stimulated our interest in the determination of the dissociation inhibition constant (K_i) for the binding of both compounds to b. β-trypsin and p. p. β-kallikrein-B, generally taken as models for serine proteinases acting on cationic substrates, and in the investigation of molecular bases for the formation of enzyme:inhibitor binary adducts (23).

For both enzyme/inhibitor systems considered, TAPP and benzamidine inhibition behaviours are strictly competitive. Moreover, titrations conform to simple equilibria, and K_i values are independent of the proteinase concentration.

The data shown in Figure 2 indicate that TAPP and benzamidine have the same affinity (i.e., K_i values) for b. β-trypsin, over the whole pH range explored (from pH 2.0 and 8.0), at 37°C. In contrast, the values of K_i for the binding of TAPP to p. p. β-kallikrein-B are lower than those for benzamidine association tending to converge at low pH. Decreasing pH from 5.5 to 2.5, the increase in the values of K_i for benzamidine binding to b. β-trypsin and p. p. β-kallikrein-B as well as for TAPP association to b. β-trypsin reflects, according to linkage relations (24), the acid pK_a shift of a single ionizing group on inhibitor binding from 4.5, in the free enzymes, to 3.8, in the proteinase:inhibitor binary adducts. In the case of TAPP binding to p. p. β-kallikrein-B the situation is different; in fact, the pH dependence of K_i for TAPP association depends on the acidic pK_a shift of two equivalent TAPP-linked proton binding groups. Such residues change their pK_a values from 4.5, in the free p. p. β-kallikrein-B, to 3.6, in the enzyme:TAPP binary complex.

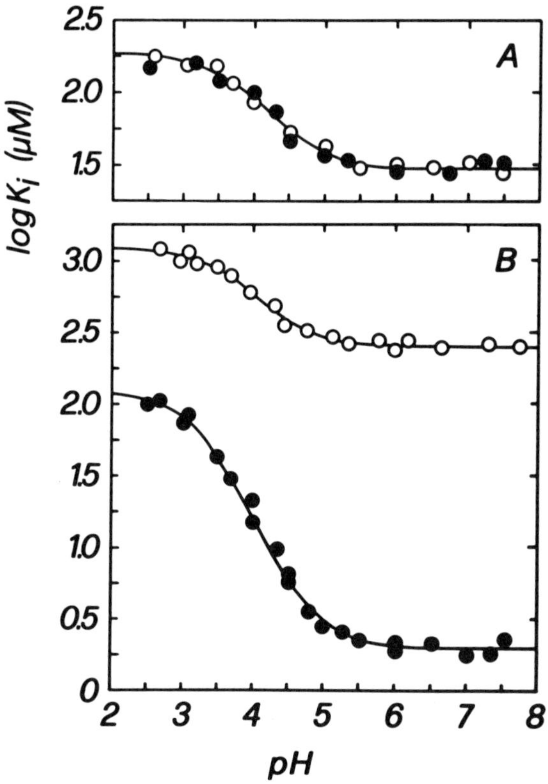

Figure 2. pH dependence of the dissociation inhibition constant (K_i) for benzamidine (○) and TAPP (●) binding to bovine β-trypsin (A) and porcine pancreatic β-kallikrein-B (B), at 37°C. The continuous lines fitting the pH dependence of K_i for benzamidine and TAPP binding to b. β-trypsin (A) as well as for benzamidine association to p. p. β-kallikrein-B (B) reflect the acidic pK_a shift of a single ionizing group from 4.5, in the free enzyme, to 3.8, in the enzyme:inhibitor binary adducts. The continuous line fitting the pH dependence of K_i for TAPP association to p. p. β-kallikrein-B (B) reflects the acidic pK_a shift of two equivalent ionizing groups from 4.5, in the free proteinase, to 3.6, in the enzyme: TAPP binary complex.

Morever, the data shown in Figure 2 indicate that the positively charged TAPP and benzamidine inhibitors bind preferentially the unprotonated (i.e., the negatively charged) species of both proteinases considered.

The pK_a values of the group(s) involved in TAPP and benzamidine binding are closely similar for b. β-trypsin and p. p. β-kallikrein-B, and agree very well with the pK_a values of amino acid residue(s) modulating, between pH 2.5 and 5.5, not only inhibitor binding but also spectral and catalytic properties of serine proteinases acting on cationic substrates (1,2,13-15).

These findings can be interpreted on the basis of the threedimensional models of b. β-trypsin:benzamidine and p. p. β-kallikrein-B:benzamidine binary adducts (3,6), as well as by means of computer graphics techniques which allowed the establishment of binding modes for TAPP to both proteinases considered (23). Thus, the ionizable group affecting benzamidine binding can be identified as the invariant Asp 189 residue present in the S_1 subsite of b. β-trypsin and p. p. β-kallikrein-B. In addition, TAPP binding to b. β-trypsin is probably modulated by the acid-base equilibrium of the Asp 189 side chain. The analysis of computer generated models of enzyme:TAPP binary adducts allowed the tentative assignment of the second ionizing group affecting TAPP binding to p. p. β-kallikrein-B. Thus, only Asp 148 as well as Asp 174 in the latter proteinase meet the charge and steric requirements for the binding of a second TAPP substituent, once the first benzamidine group is bound in the S_1 subsite of the enzyme, which interacts with the Asp 189 residue. Although this stereochemical interpretation agrees with the pH independence of K_i for benzamidine and TAPP binding to bovine α-chymotrypsin (20), where aspartyl and/or glutamyl residue(s) affecting inhibitor association have no counterpart(s) (25), other subtle structural sources of enzyme:inhibitor complexes (de)stabilization cannot be ruled out.

As a whole, although enzyme:inhibitor affinity cannot be directly related to stereochemical considerations, multiple interactions, only occurring in the p. p. β-kallikrein-B:TAPP binary adduct, may be taken as indicative of the higher inhibitor effect shown by this aromatic tetra-benzamidine derivatives as compared to that observed for benzamidine. Moreover, the enzyme: TAPP system may be a valuable model for the study of protein:protein interactions.

CONCLUSIONS

The recognition between proteinases and their respective low molecular weight ligands (i.e., substrates and inhibitors) is based on stereospecific grounds. In fact, the interface between the enzyme and its ligand at the level of the primary specificity subsite is stabilized by close packing of atoms, pairing of oppositely charged groups and complementarity of hydrophobic regions. However, the contact between the two molecules is not static but is accompanied by transient structural changes that maximize the interactions. In general, the functional perturbation or small specificity variations arise from local alterations of amino acid residues. On the other hand, the phenomenological behaviour of serine proteinases may be insensitive to the replacement of some amino acid residues in the specificity cleft if the ligand is sufficiently flexible (e.g., lysine *versus* arginine).

REFERENCES

1. Ascenzi, P., Menegatti, E., Guarneri, M., Bolognesi, M. and Amiconi, G. (1984) Biochim. Biophys. Acta 789, 99-103.
2. Ascenzi, P. , Menegatti, E., Bolognesi, M., Guarneri, M. and Amiconi, G. (1986) Biochim. Biophys. Acta 871, 319-323.

3. Bode, W. and Schwager, P. (1975) J. Mol. Biol. 98, 693-717.
4. Furie, B., Bing, D.H., Feldmann, R.J., Robison, D.J., Burnier, J.P. and Furie, B.C. (1982) J. Biol. Chem. 257, 3875-3882.
5. Huber, R. and Bennett, W.S. Jr. (1983) Biopolymers 22, 261-279.
6. Bode, W., Chen, Z., Bartels, K., Kutzback, C., Schmidt-Kastner, G. and Bartunik, H. (1983) J. Mol. Biol. 164, 237-282.
7. Chen, Z. and Bode, W. (1983) J. Mol. Biol. 164, 283-311.
8. Steitz, T.A. and Shulman, R.G. (1982) Ann. Rev. Biophys. Bioeng. 11, 419-444.
9. Peller, L. and Alberty, R.A. (1959) J. Am. Chem. Soc. 81, 5907-5914.
10. Menegatti, E., Scalia, S., Bortolotti, F., Ascenzi, P., De Marco, A. and Guarneri, M. (1985) Pharm. Acta Helv. 60, 170-174.
11. Menegatti, E., Bolognesi, M., Scalia, S., Bortolotti, F., Guarneri, M. and Ascenzi, P. (1986) J. Pharm. Sci. (in the press).
12. Jesty, J. and Nemerson, Y. (1976) Methods Enzymol. 45, 95-107.
13. Antonini, E., Ascenzi, P., Menegatti, E., Bortolotti, F. and Guarneri, M. (1982) Biochemistry 21, 2477-2482.
14. Ascenzi, P., Menegatti, E., Guarneri, M., Bortolotti, F. and Antonini, E. (1982) Biochemistry 21, 2483-2490.
15. Ascenzi, P., Torroni, A., Menegatti, E., Guarneri, M. and Amiconi, G. (1985) Biochim. Biophys. Acta 832, 215-219.
16. Kasai, S., Arimura, H., Nishida, M. and Suyama, T. (1985) J. Biol. Chem. 260, 12382-12389.
17. Castellino, F.J. and Powell, J.R. (1981) Methods Enzymol. 80, 365-378.
18. Lottspeich, F., Geiger, R., Henschen, A. and Kutzback, C. (1979) Hoppe-Seyler's Z. Physiol. Chem. 360, 1947-1950.

19. Parsons, D.G. (1972) Minnesota 3M Laboratories, Brit. Pat. 1, 288, 376, 6 Sept. 1972, C.A. 78, 3980r.
20. Menegatti, E., Guarneri, M., Ferroni, R., Bolognesi, M., Ascenzi, P. and Antonini, E. (1982) FEBS Letters 141, 32-36.
21. Ferroni, R., Menegatti, E., Guarneri, M., Taddeo, U., Bolognesi, M., Ascenzi, P. and Amiconi, G. (1984) Il Farmaco - Ed. Sc. 39, 901-909.
22. Ferroni, R., Menegatti, E., Orlandini, P., Guarneri, M., Taddeo, U., Bolognesi, M., Ascenzi, P., Bertollini, A. and Amiconi, G. (1986) Il Farmaco - Ed. Sc. 41, 464-470.
23. Menegatti, E., Ferroni, R., Scalia, S., Guarneri, M., Bolognesi, M., Ascenzi, P. and Amiconi, G. (1987) J. Enzyme Inhibition (in the press).
24. Wyman, J. (1964) Adv. Protein Chem. 19, 223-286.
25. Tsukada, H. and Blow, D.M. (1985) J. Mol. Biol. 184, 703-711.

INTERACTION BETWEEN SERINE (PRO)ENZYMES AND MACROMOLECULAR INHIBITORS. THERMODYNAMIC AND KINETIC ASPECTS

Gino Amiconi[≠], Paolo Ascenzi[≠], Martino Bolognesi[ø], Enea Menegatti[+] and Mario Guarneri[+]

[≠]C.N.R., Center for Molecular Biology, Department of Biochemical Sciences, University of Rome "La Sapienza", 00185 Rome, Italy

[ø]Department of Genetics and Microbiology, Section of Crystallography, University of Pavia, 27100 Pavia, Italy

[+]Department of Pharmaceutical Sciences, University of Ferrara, 44100 Ferrara, Italy

Recognition of a macromolecular substrate or inhibitor by an enzyme is not a single-event process, but consists in a spatio-temporal series of interactions. Therefore, examination of the final binding product evidenced in the X-ray crystallography (that is in a time-averaged structure) does not fully define all the reactivity characteristics exhibited by the system. In particular, the interaction of the protein surfaces is a complex event involving molecular flexibility, induced fit, entropy changes, solvent, hydrophobic, van der Waals and electrostatic interactions. Therefore, for a given interacting system it is highly desirable to have as much information as possible from thermodynamic and kinetic data obtained in various states of the functional events. Some questions can illustrate the kind of information sought. How does each residue at the interface contribute to the energetics of the complex stabilization? Are the pairs of interactions at the contact surface independent of the rest of the molecule, i.e., is the contact like that between two rigid surfaces? Or are the interactions modulated by changes in conformation that accompany adduct formation i.e., does flexibility serve as a regulatory property? Is the interaction dynamics better described with a zipper-like model or with a groping movement?

The strategies and processes here discussed are restricted to the formation of non-covalent complexes between serine (pro)enzymes and macromolecular inhibitors.

DETERMINATION OF THERMODYNAMIC AND KINETIC PARAMETERS FOR BINARY COMPLEX FORMATION

Values of thermodynamic parameters for the binding of bovine (b.) and porcine (p.) pancreatic secretory trypsin inhibitor (PSTI; Kazal-type inhibitor) to the serine (pro)enzymes considered have been obtained: (i) by the evaluation of the inhibitory effect of b. and p. PSTI on the bovine (b.) β-trypsin catalyzed hydrolysis of N-α-carbobenzoxy-L-arginine p-nitrophenyl ester (ZArgONp) as well as on the bovine (b.) α-chymotrypsin catalyzed hydrolysis of N-α-carbobenzoxy-L-tyrosine p-nitrophenyl ester (ZTyrONp); and (ii) by following the spectral changes associated to the formation of the bovine (b.) trypsinogen:b. or :p. PSTI adducts (1).

Values of kinetic parameters for b. and p. PSTI binding to b. β-trypsin have been obtained by following spectrophotometrically: (i) the displacement of proflavine or benzamidine from the central site of the enzyme upon inhibitor binding; (ii) the intrinsic optical density changes in the ultraviolet region accompanying the enzyme:inhibitor adduct formation; and (iii) the loss of the enzyme activity (on the hydrolysis of ZArgONp) associated to inhibitor binding (2-4).

THERMODYNAMICS OF THE BINDING OF KAZAL- AND KUNITZ-TYPE INHIBITORS TO SERINE (PRO)ENZYMES

The information for the determination of the proper structure of the interaction product between two proteins can be assumed to be all contained in the macromolecules themselves. Hence the interaction mechanism at the basis of the serine (pro)enzyme:inhibitor adduct formation may be elucidated in vitro by mixing the purified components since no additional device or regulatory process is needed.

Under all the experimental conditions used, the binding of b. and p. PSTI to b. β-trypsin, b. α-chymotrypsin and b. trypsinogen conforms apparently to simple equilibria (see Fig. 1).

Table 1 shows values of K_a, $\Delta G°$ and $\Delta S°$ (at 21°C) as well as of $\Delta H°$ (obtained between 5°C and 45°C) for various binding reactions.

As reported in Fig. 1 and Table 1, b. PSTI shows a systematically higher affinity for the three serine (pro)enzymes considered as compared to p. PSTI. Under the same experimental conditions, the affinity of the bovine basic pancreatic secretory trypsin inhibitor (BPTI; Kunitz-type inhibitor) for the serine (pro)enzymes considered is higher by at least two orders of magnitude than that observed for the binding of Kazal-type inhibitors.

The higher affinity of BPTI for the three serine (pro)enzymes as compared to b. and p. PSTI cannot be related in simple terms to the number of hydrogen bonds and van der Waals interactions at the contact surfaces. For instance, binding of BPTI to b. trypsinogen involves a lower number of interactions than p. PSTI. However, upon BPTI binding, b. trypsinogen adopts a structure that is almost coincident with that of b. β-trypsin (5), whereas, p. PSTI induces in the zymogen a novel conformation at the activation domain that differs from that of b. β-trypsin (6) and thus may be energetically less favourable. This approach could be applied to each serine (pro)enzyme studied and each inhibitor. It may be considered that some of the intrinsic binding energy between the two proteins is expended to order a mobile region of the macromolecule(s), so that the net binding strength will be weaker than the sum of the interactions would suggest. This is certainly true in the case of b. trypsinogen and b. β-trypsin, which should have the same number and kind of interactions with each inhibitor, but show very different affinity constants, i.e., identical interactions associated with different binding energy. Hence flexible segments linked to the active center of the serine (pro)enzyme appear to contribute directly to the optimization

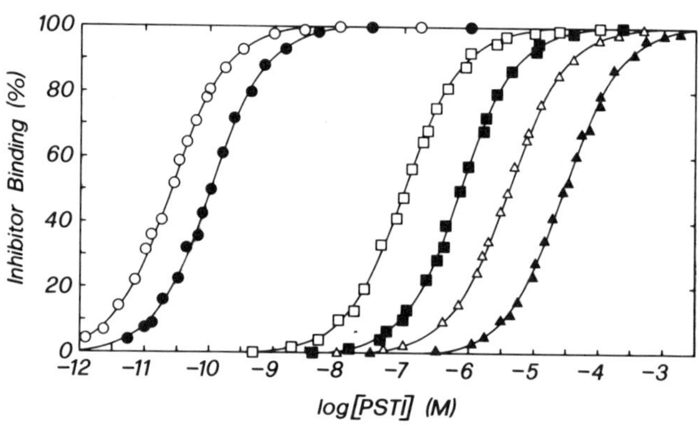

Fig. 1. Amount of bovine PSTI (open symbols) and porcine PSTI (filled symbols) bound to bovine β-trypsin (circles), bovine α-chymotrypsin (squares) and bovine trypsinogen (triangles) *versus* the free inhibitor concentration, at pH 8.0 and 21°C. Continuous lines are the optimal fits calculated on the basis of the single binding-site model, allowing the determination of Ka values given in Table 1.

TABLE 1
Thermodynamic Parameters for the Binding of Bovine and Porcine PSTI to Serine (Pro)enzymes (pH=8.0)

Complex	K_a M^{-1}	ΔG° kcal/mol	ΔH° kcal/mol	ΔS° e.u.
β-trypsin:b.PSTI[a]	4.0×10^{10}	−14.2	0.0	+48
β-trypsin:b.PSTI[b]	3.1×10^{10}	−14.4	−1.4	+44
α-chymotrypsin:b.PSTI[a]	1.0×10^{7}	−9.4	+1.8	+38
trypsinogen:b. PSTI[a]	2.5×10^{5}	−7.2	+2.5	+33
β-trypsin:p. PSTI[a]	1.0×10^{10}	−13.4	+0.5	+47
α-chymotrypsin:p. PSTI[a]	1.4×10^{6}	−8.2	+2.2	+35
trypsinogen:p. PSTI[a]	3.5×10^{4}	−6.1	+3.1	+31

[a]Values of thermodynamic parameters determined from equilibrium measurements. Values of K_a, ΔG° and ΔS° have been obtained at 21°C; values of ΔH° have been determined between 5°C and 45°C.

[b]Values of thermodynamic parameters calculated from kinetic experiments. Values of K_a, ΔG° and ΔS° have been obtained at 25°C; the value of ΔH° was obtained between 1°C and 37°C.

of its interactions with the inhibitor (or the substrate) and therefore to the specificity of the molecular recognition-transformation process. As far as b. α-chymotrypsin is concerned, its lower affinity for all inhibitors with respect to b. β-trypsin finds a partial explanation in the absence of a negative charge at the bottom of its specificity pocket with the consequent absence of the salt-bridge with Lys 15 (of BPTI) or Arg 18 (of b. PSTI) or Lys 18 (of p. PSTI).

The differences observed in the affinity of b. and p. PSTI for b. β-trypsin, b. α-chymotrypsin and b. trypsinogen (see Fig. 1 and Table 1) can be related to amino acid substitutions occurring at positions 11 and 14 of Kazal-type inhibitors (Ser in p. PSTI and Asn in b. PSTI). The presence of Asn residues at the (pro)enzyme:inhibitor (re)active region may allow additional protein-protein interactions which should stabilize the (pro)enzyme:b. PSTI adduct.

Both the $\Delta H°$ and $\Delta S°$ values for b. and p. PSTI binding to the three (pro)enzymes are comparable with those reported for BPTI association to serine proteinases and their zymogens (7-11), and indicate that the association with the inhibitor is an entropy-driven process. It is well known that electrostatic bonding and hydrophobic interactions are usually accompanied by the release of constrained water molecules that give a positive contribution to the observed $\Delta S°$ and may give positive $\Delta H°$ terms. Conformational changes in proteins can also give large and unpredictable changes in thermodynamic parameters due to changes in solvent interactions, as well as to the complex structure of the protein itself. Even though it is well appreciated how easily the compensation of $\Delta H°$ and $\Delta S°$ can lead to misinterpretation of observed thermodynamic parameters, the experimental results on the (pro)enzyme:inhibitor systems are unequivocal.

As a whole these data indicate that single amino acid substitutions (e.g., b. PSTI <u>versus</u> p. PSTI) occurring at the (pro)enzyme:inhibitor (re)active sites impose small perturbations in the binding energy, which affect the equilibrium constants between (pro)enzyme and inhibitor (or substrate) and may influence the activation energy of the chemical step. Evidence for the use of binding energy in catalysis has come indirectly from studies on the binding of modified substrates or specifically designed inhibitors as transition state analogues (12). Therefore, recognition, based on complementarity between two surfaces, should be considered not only from a static point of view, but also during the catalytic event; in fact the enzyme may take advantage of the geometrical changes both in the (pro)enzyme molecule itself and in the substrate (or inhibitor) to bind the latter more tightly at some particular point along the entire reaction profile.

KINETICS OF THE BINDING OF KAZAL- AND KUNITZ-TYPE INHIBITORS TO BOVINE β-TRYPSIN

Four different signals, namely the displacement of proflavine and benzamidine, the optical density changes in the ultraviolet region, and the loss of the enzymatic activity can be employed to follow the time-course of the inhibitor binding reaction. Under conditions where inhibitors are at least in five fold excess over the enzyme, the reaction of b. and p. PSTI as well as BPTI with b. β-trypsin corresponds to a pseudo-first-order process (2,4,13,14).

For the interaction of the three macromolecular inhibitors with b. β-trypsin, the concentration dependence of the rate(s) is second-order at low inhibitor concentrations and tends to become first-order at high concentrations (see Fig. 2). This behaviour has been interpreted (2,4,13,14) as indicative of the presence of a relatively fast pre-equilibrium followed by a limiting first-order process, according to the scheme:

$$E + I \overset{K_i}{\rightleftharpoons} (E{:}I)_1 \xrightarrow{k_{+i}} (E{:}I)_2 \qquad (I)$$

where $(E{:}I)_1$ is a labile, rapidly formed complex; $(E{:}I)_2$ is the final stable adduct; K_i is the pre-equilibrium constant; and k_{+i} is the limiting pseudo-first-order rate-constant.

Figure 2 shows the dependence of the pseudo-first--order rate-constant around neutrality and at low pH, as measured by the four different methods. Where the comparison is possible, the values of the kinetic parameters for the b. β-trypsin:b. PSTI adduct formation, agree well with one another, both at pH 7.5 and at pH 3.5. On the other hand, in the case of the b. β-trypsin : p. PSTI complex formation, the results obtained by the four different signals agree with one another only at neutral pH, but differ widely at low pH. Such a pH dependent effect has been reported first for the b.β-trypsin:BPTI adduct formation (2,14).

The analysis of these data allows two relevant considerations. (i) Binding of b. PSTI to b.β-trypsin may be regarded as a simple two-state event, a view confirmed also by the good agreement between the values of the thermodynamic parameters determined directly (present paper) and those calculated from kinetic quantities (8) (see Table 1). (ii) In contrast, the association of p. PSTI to b. β-trypsin may be described at high temporal resolution as a multi-step process (14) involving several intermediates with distinct functional properties. In fact, at pH 3.5, upon p. PSTI binding, the displacement of proflavine (a competitive inhibitor) from b. β-trypsin precedes the intrinsic optical density changes in the ultraviolet region, and the latter precedes the loss of the enzyme activity as well as the displacement of benzamidine from the proteinase (see Fig. 2). These data lead to the prediction that, at pH 3.5, p. PSTI addition to the b. β-trypsin: proflavine complex should initially remove proflavine with the concomitant transient recovery of the catalytic activity, while the irreversible inhibition should take place only upon completion of p. PSTI binding. The kinetic data shown in Fig. 3 verify this prediction, indicating that during the formation of the b. β-trypsin: p. PSTI complex (at least) one transient intermediate is formed which is not able to interact with proflavine,

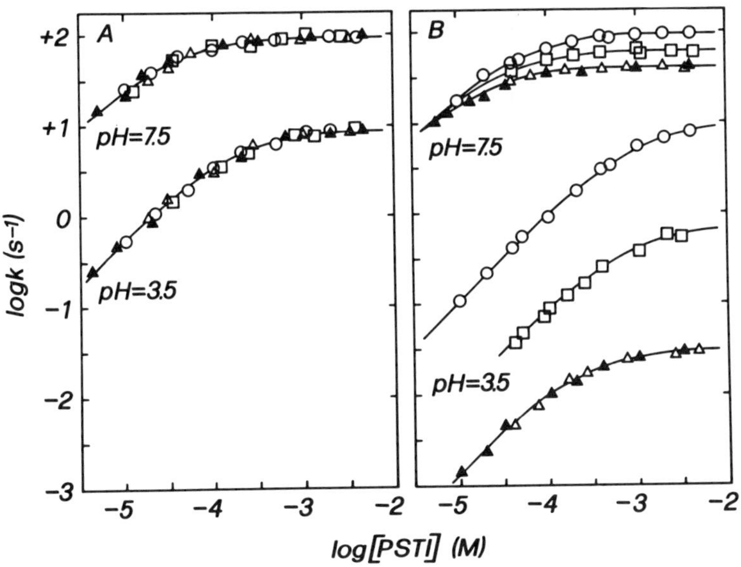

Fig. 2. Dependence of the pseudo-first-order rate-constant (k) for the binding of bovine and porcine PSTI (panels A and B, respectively) to bovine β-trypsin on the inhibitor concentration for (i) proflavine displacement (○); (ii) intrinsic spectral changes in the ultraviolet region (□); (iii) benzamidine displacement (△); and (iv) loss of the enzymatic activity (▲), at pH 7.5 (top) and pH 3.5 (bottom); T=21°C. The continuous lines are theoretical curves calculated on the basis of the model shown in Scheme I, taking into account a relatively fast pre-equilibrium followed by a limiting first-order process.

but may bind and hydrolyze the substrate (namely, ZArgONp). Thus, when p. PSTI is added to a solution containing b. β-trypsin, proflavine and ZArgONp, the rate of hydrolysis of the substrate increases to a value corresponding to that observed in the absence of

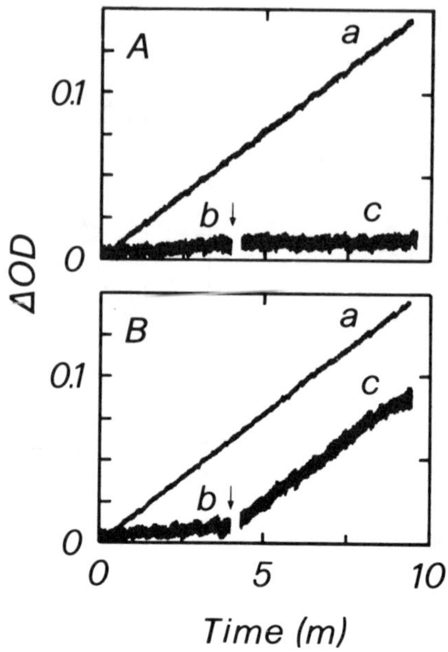

Fig. 3. Effect of bovine and porcine PSTI (panels A and B, respectively) on the time-course of the bovine β-trypsin catalyzed hydrolysis of ZArgONp at pH 3.5 and 21°C. Traces a in the absence of macromolecular inhibitors and proflavine; traces b in the presence of 1 mM proflavine; traces c as b but after the addition of b. or p. PSTI (panels A and B, respectively), indicated by the arrows. B. β-trypsin concentration, 0.1 μM; ZArgONp concentration, 85 μM; inhibitor concentration, 20 μM; λ = 360 nm.

proflavine. Such a behaviour, known as chronosteric effect, has been observed for the first time in the formation of the b. β-trypsin:BPTI adduct (3). As expected from the data shown in Fig. 2, the chronosteric effect does not occur in the association of b. PSTI with b. β-trypsin (see Fig. 3).

The reported differences in the kinetic properties concerning the formation of the adducts of b. β-trypsin with b. and p. PSTI as well as BPTI are likely to be related to the substitution of single amino acid residues at the (re)active sites of inhibitors as well as of (pro)enzymes. In this respect, the presence of Asn residues at positions 11 and 14 in b. PSTI (Ser in p. PSTI) may allow different enzyme:inhibitor interactions, which may be reflected in the association mechanism. During casual collisions, a long persistence of protein-protein association has been shown: the protein molecules in these complexes are free to rotate over each other's surface, allowing an efficient search for (re)active surface regions in the course of a single protein collision (15). Since the reactive centre is small compared to the total surface area of the interacting macromolecules, the protein recognition process could be simply visualized as a pattern recognition, i.e., as the search during the lengthy collision of a pair of complementary topographycal patterns which minimize the free-energy of association. However, this model fails to dock correctly BPTI onto b. β-trypsin by a computational method (16) in line with the experimental evidence that in such a system the chronosteric effect is present (3). Therefore, strong interactions (such as those between serine (pro)enzymes and their macromolecular inhibitors) appear to require not only a large number of interacting groups appropriately situated over sufficiently large surfaces of both proteins, but also "something else", possibly something like a dynamic complementarity (17).

In conclusion, the presence of the chronosteric effect (e.g., in the complex formation between p. PSTI or BPTI and b. β-trypsin) may be taken as indicative of a conformational adjustment of protein inhibitors to their target (pro)enzymes. On the other hand, in the case of the association of b. β-trypsin with b. PSTI (a process without chronosteric effect) in which the inhibitor shows only two substitutions in the reactive centre compared to p. PSTI, recognition could in principle be attributed to a perfect static complementarity, i.e. to a purely geometrical docking, but probably the event is also somehow regulated by a dynamic matching of the interacting surfaces.

CONCLUSIONS

The crystallographic analysis of b.β-trypsin:macromolecular inhibitor complexes: indicates the existence of multiple points of attachment; proves a rigorous spatial configuration; and, in the adducts, attests the presence of many energetically equivalent contributions. The fact that the binding energy depends upon many small, apparently independent contributions has some important consequences: (i) since no atomic group in the enzyme or inhibitor is all important, it follows that macromolecular inhibitors differing from one another by a single feature, like the replacement of one or two group(s), will in many cases be bound as well; (ii) if protein specificity of binding is the result of many independent effects none of which is dominant, the resulting property (i.e., the binding energy) will have a normal distribution with respect to those structural changes in the protein that give different weights to the individual interactions; (iii) on this basis, the very origin of molecular recognition implies that it cannot be perfect in any sense.

These conclusions are of general relevance in enzymology, in that catalysis itself appears to be an event "delocalized" around the active site; in other words catalysis results not from just a few key-residues but is the consequence of many interactions of small binding energy, that conspire to easy the reagents into the transition state and then stabilize the high energy intermediate (18,19). In addition flexible segments, linked to the recognition site, can affect the binding energy by taking advantage of entropy loss; in other words, the system may have high specificity without excessively tight binding (20).

From a kinetic point of view, the initial collision (with the consequent displacement of proflavine) has a substantial duration (related to optical density change in the ultraviolet region) in which the molecules are free to explore each other's surfaces. However, the movement of the two interacting surfaces is not a smooth steering towards a sequential formation of hydrogen bonds, polar and apolar interactions; during the process, the amino acids relevant to the binding bindly grope along, randomly breaking a salt bridge here, forming a hydrogen bond there, untill the system nestles in an energy minimum along the reaction pathway (monitored by the loss of enzymatic activity and the displacement of benzamidine).

REFERENCES

1. Antonini, E., Ascenzi, P., Bolognesi, M., Gatti, G., Guarneri, M. and Menegatti, E. (1983) J. Mol. Biol. 165, 543-558.
2. Antonini, E., Ascenzi, P., Menegatti, E. and Guarneri, M. (1983) Biopolymers 22, 363-375.
3. Antonini, E., Ascenzi, P., Bolognesi, M., Menegatti, E. and Guarneri, M. (1983) J. Biol. Chem. 258, 4676-4678.
4. Ascenzi, P., Bolognesi, M., Guarneri, M., Menegatti, E. and Amiconi, G. (1984) Mol. Cell. Biochem. 64, 139-144.
5. Bode, W., Schwager, P. and Huber, R. (1978) J. Mol. Biol. 118, 99-112.
6. Bolognesi, M., Gatti, G., Menegatti, E., Guarneri, M., Marquart, M., Papamokos, E. and Huber, R. (1982) J. Mol. Biol. 162, 839-868.
7. Vincent, J.P. and Lazdunski, M. (1973) Eur. J. Bio-

chem. 38, 365-372.

8. Schweitz, H., Vincent, J.P. and Laskowski, M. (1973) Biochemistry 15, 2841-2846.
9. Vincent, J.P. and Laskowski, M. (1976) FEBS Letters 63, 240-244.
10. Menegatti, E., Guarneri, M., Bolognesi, M., Ascenzi, P. and Amiconi, G. (1984) J. Mol. Biol. 176, 425-430.
11. Menegatti, E., Guarneri, M., Bolognesi, M., Ascenzi, P. and Amiconi, G. (1986) J. Mol. Biol. 191, 295-297.
12. Fersht, A.R. (1985) Enzyme Structure and Mechanism Freeman, San Francisco.
13. Ascenzi, P., Amiconi, G., Bolognesi, M., Menegatti, E. and Guarneri, M. (1986) Biopolymers 25, 2325-2333.
14. Menegatti, E., Guarneri, M., Bolognesi, M., Ascenzi, P. and Antonini, E. (1983) Il Farmaco - Ed. Sc. 38, 978-997.
15. Sommer, J., Jonah, C., Fukuda, R. and Bersohn, R. (1982) J. Mol. Biol. 159, 721-744.
16. Connolly, M.L. (1986) Biopolymers 25, 1229-1247.
17. Lumry, R. and Gregory, R.B. (1986) in: The Fluctuating Enzyme (Welch, G.R. ed.) pp. 1-190, Wiley, New York.
18. Wells, T.N.C. and Fersht, A.R. (1986) Biochemistry 25, 1881-1886.
19. Ho, C.K. and Fersht, A.R. (1986) Biochemistry 25, 1891-1897.
20. Ringe, D. and Petsko, G.A. (1985) Prog. Biophys. Mol. Biol., 45, 197-235.

SUBUNIT INTERACTIONS IN CYTOCHROME OXIDASE:

THE ROLE OF SUBUNIT III

P. Vecchini^, G. Antonini*, F. Malatesta*,
P. Sarti^, M. Wilson°, and M. Brunori^

^Dept. of Biochemical Sciences and Center of Molecular Biology (CNR), Univ. of Rome "La Sapienza", Italy; *Dept. of Experimental Medicine and Biochemical Sciences, Univ. of Rome "Tor Vergata," Italy; °Dept. of Chemistry, Univ. of Essex, U.K.

Cytochrome c oxidase is an oligomeric membrane protein that catalyzes the oxidation of cytochrome c and the reduction of oxygen to water. The enzyme-mediated proton translocation is also known to be linked to its electron transfer activity either in mitochondria or in artificial phospholipid vesicles (1-4). The protein has been purified from several sources with different procedures, resulting in a variable number of co-purifying subunits (7-12) whose specific functions are still, to a large extent, obscure. According to the latest experimental findings in prokaryotes, however, enzyme activity is linked to the presence of at least the three largest subunits (I, II and III), which are coded by mitochondrial DNA in eukaryotes. Subunit I and II contain the four redox active metal centers; two of these constitute the electron accepting pole (cytochrome a, Cu a), where electron transfer to-and-from cytochrome c takes place; the other two form a heterobinuclear center (cytochrome a_3, Cu a_3), which receives electrons delivered from the former sites and where both binding and reduction of dioxygen take place (for reviews see 5,6). Subunit III is among the most hydrophobic polypeptides within the enzyme, and although it does not bear metals, it is known to affect the efficiency of proton pumping (4).

It has proved possible to isolate beef heart cytochrome oxidase in two-dimensional crystals suitable for electron microscopy studies and image reconstruction. A low resolution structure for the protein has thus been obtained (7), showing that cytochrome oxidase has the shape of a Y. Two different crystal types have been obtained using non-ionic detergents and deoxycholate: in the former case the protein appears as a dimer, whereas with the latter detergent the dimer dissociates into monomers (8,9). Both types of aggregates appear to be enzymatically competent. For the purpose of the present paper we will define as one functional unit of cytochrome oxidase (i.e. a single monomer) the molecular unit containing the four redox active metal centers and the full set of subunits (twelve polypeptides) in a 1:1 stoichiometry (with the exception of subunit VIIIB which is present twice). The molecular weight of this enzymatically competent oligomeric unit is calculated to be 200,000 daltons. The native enzyme, as solubilized by detergents such as Triton X 100 (Tx 100), Tween 80 or sodium cholate has been shown to exist in different aggregation states (6). The basic Yonetani procedure to purify the enzyme from ox heart (10) yields a protein containing a high percentage of dimers (i.e. two functional units); on the other hand, the enzyme purified from rat liver, camel and shark heart is in a monomeric state (11).

The factors affecting quaternary interactions in cytochrome oxidase are without a doubt manyfold and complex. Some of the relevant parameters which must be controlled are: pH, ionic strength, detergent type and amount, protein concentration and polypeptide composition.

In the present study we report hydrodynamic measurements of detergent-solubilized cytochrome oxidase under varying conditions. We have devised an empirical procedure to assess the homogeneity of a given enzyme preparation, demonstrating moreover the existence of multiple equilibria between different aggregation states of cytochrome oxidase. These equilibria were found to be modulated by a number of environmental factors. Finally, evidence is herein reported suggesting that the redox state of the detergent solubilized enzyme does not grossly perturb the protein subunits recognition and therefore its quaternary structure.

MATERIALS AND METHODS

Beef heart cytochrome c oxidase (E.C. 1.9.3.1) was purified according to Yonetani (10) and dissolved in either 0.1 M Na-phosphate, 0.2% cholate pH 7.4 or 0.1 M K-Hepes buffer containing the same detergent. Prior to ultracentrifugation the protein was either diluted directly in the desired buffer or dialyzed. Protein concentration was determined using an extinction coefficient (reduced-oxidized) of 11 mM^{-1} cm^{-1} (6) at 605 nm. Subunit III-depleted cytochrome c oxidase was prepared as described previously (12). All preparations subjected to ultracentrifugation analysis were enzymatically active as determined by parallel stopped-flow measurements using horse heart ferrocytochrome c as substrate. Sedimentation velocity experiments were carried out using a Spinco Model E analytical ultracentrifuge equipped with a temperature control unit. Unless otherwise stated, samples were centrifuged at 52,000 rpm at 5-10°C. Schlieren photographs were taken at the time indicated in the Figure legends after reaching the pre-set speed. Sedimentation coefficients (S20) were corrected only for temperature (20°C) because of the detergents present in the buffer. Cytochrome oxidase-containing vesicles (COV) were prepared according to Sarti et al. (13) and respiratory control ratios (RCR) were determined as the ratio of cytochrome oxidase activity in the presence of ionophores (valinomycin and CCCP) to that in their absence either polarographically or spectrophotometrically.

RESULTS AND DISCUSSION

Determinations of the molecular weight of beef heart cytochrome oxidase in non-ionic detergents at physiological pH indicate that the stable species from this source is a dimer. Under similar conditions cytochrome oxidase extracted from shark heart (and other elasmobranchs) is more stable as a monomer (14). The reason for this behavior is presently not clear, but one explanation can be found in the interspecies heterogeneity of the polypeptide composition. Although the polypeptide composition of cytochrome oxidase is quite similar (as seen, for example, from the immunological cross-reactivity) the local chemical composition may be significantly different at quaternary sites responsible for the stabilization of a given aggregate structure. Therefore

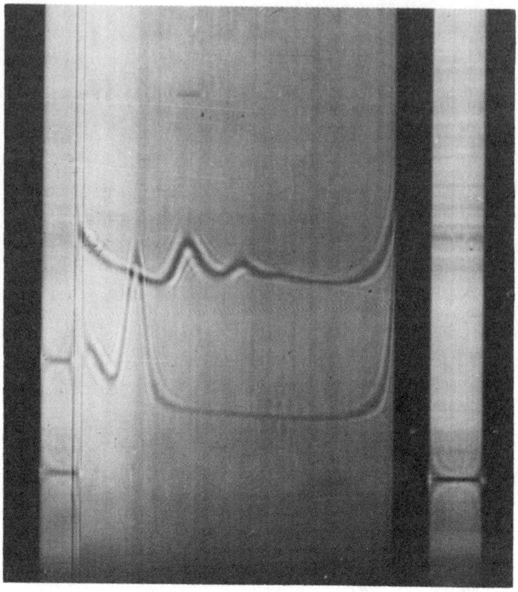

Figure 1. Sedimentation velocity patterns of cytochrome c oxidase. Buffer: 86 mM phosphate, pH 7.3. Upper: native enzyme in the presence of 0.5% Tween 80. Lower: III-less oxidase in the presence of 0.5% Tx 100. Photograph taken 31 min. after reaching full speed of 51,000 rpm. For details see text.

one can conclude that the polypeptide composition may condition the aggregation state of cytochrome oxidase. This is clearly shown in Figure 1.

The lower Schlieren pattern represents cytochrome oxidase from which subunit III has been removed (together with smaller polypeptides) by treatment with α-chymotrypsin followed by ion exchange chromatography on DEAE 52 in Triton X 100. The sedimentation velocity profile shows the presence of a symmetrical peak with $s_{20}= 6$. This preparation is quite monodisperse and corresponds to a monomer (one functional unit) (15). The upper Schlieren profile represents the enzyme carrying the full set of subunits under the same conditions of buffer and pH but in the presence of 0.5% Tween 80. The s_{20}

value for the main peak (75%) is equal to 12-13, consistently with the values reported in the literature for dimers (15). The monomer peak shape and position is independent of the redox state of the enzyme; in fact, the III-less enzyme, after dithionite reduction and exposure to saturating carbon monoxide shows an s_{20} = 5.7, a value close to the one pertaining to the fully oxidized III-less protein.

This result implies that, at least to some extent, subunit III is involved in the stabilization of the dimer of cytochrome c oxidase. However, polypeptide composition is not the only factor controlling quaternary interactions. The type of detergent used to disperse cytochrome oxidase and its amount relative to the protein also influence which aggregate will be stabilized. This is clearly shown in Table 1.

Table 1. Correlation between aggregation state and respiratory control ratios (RCR).

Enzyme	Detergent	%P	%D	%M	●RCR
Native	Cholate 0.2%	90	10	0	5.0
Native	Tween 80 0.5%	20	■75	5	6.5
Native	Tx 100 0.5%	0	50	50	9.5
III-less	Tx 100 0.5%	0	5	*95	3.4

P = polymer (s_{20}>20) D = dimer (s_{20} = 12-13) M = monomer (s_{20} = 8-9)
●Determined either polarographically or spectrophotometrically (see Methods section).
■8 independent determinations s_{20} = 12.1 ± 0.6
*In the case of III-less oxidase the value of s_{20} = 6.1 ± 0.4 (8 independent determinations) was measured.

Solubilization of cytochrome oxidase in 0.2% sodium cholate yields a preparation in which 90% of the total protein is in a multimeric state. Protein dispersion by Tween 80 significantly decreases the proportion of aggregated material and most of the protein sediments as a dimer. Use of the strong non-ionic detergent Tx 100 further dissociates cytochrome oxidase into single monomeric units and their relative amount increases steadily as the detergent-to-protein weight by weight ratio increases according to Robinson and Talbert (16).

The pH of the medium is also another parameter which significantly affects this multiple equilibrium, the pK of dissociation being above 10 (17, and Wilson, M.T., personal communication). Therefore, it can be concluded that the proportion of a given aggregate is quite sensitive to the type and amount of detergent used at a given pH and protein concentration. This means that changes in detergent concentration quite dramatically affect the relative proportion of aggregates. It is important, in many respects, to assess which aggregate is competent in electron transfer and in proton pumping. Cytochrome c oxidase reconstituted into artificial phospholipid vesicles with a given orientation, is a simple model system mimicking the functional properties of cytochrome oxidase in the mitochondria. To date it is not known which aggregate is functional when integrated into the bilayer. Table 1 also shows the RCR (see Materials and Methods section) of a given cytochrome oxidase sample as a function of the dispersing detergent. A clear correlation between the RCR, which reflects a proper incorporation of cytochrome oxidase in the bilayer, and detergent type could be suggested. In the experiments herein reported, the enzyme was preincubated for 30 min. at room temperature in the specified detergent, assayed for the sedimentation pattern and then reincorporated into liposomes. The data presented in Table 1 clearly indicate that the aggregation state of the native enzyme is controlled by detergent and affects the efficiency of reincorporation into vesicle (RCR). This interpretation, however, cannot be applied to the III-less enzyme, where despite a quite monodisperse state of the starting material (95% of monomers) the RCR are always below the corresponding control values. This finding possibly suggests that the subunit-subunit interactions in cytochrome oxidase can also play an important role in controlling the incorporation of the protein into artificial phospholipid membrane.

Acknowledgements. This work was partially supported by the Ministero della Pubblica Istruzione of Italy to M. Brunori and by the Progetto Strategico Biotecnologie (CNR), Italy. Mrs. Beatrice Vallone and Mr. Emilio D'Itri are gratefully acknowledged for technical assistance.

REFERENCES

1. Wikstrom, M.K.F. (1977) Nature (London) 266, 271-273.
2. Sigel, E. and Carafoli, E. (1978) Eur. J. Biochem. 89, 119-123.
3. Casey, R.P., Chappell, J.B. and Azzi, A. (1979) Biochem. J. 182, 149-156.
4. Sarti, P., Jones, M.G., Antonini, G., Malatesta, F., Colosimo, A., Wilson, M.T. and Brunori, M. (1985) Proc. Natl. Acad. Sci. USA 82, 4876-4880.
5. Wikstrom, M.K.F., Krabs, K. and Saraste, M. (1981) Cytochrome Oxidase a Synthesis, Academic Press, London.
6. Brunori, M., Antonini, G., Malatesta, F., Sarti, P. and Wilson, M.T. (1987) Adv. Inorg. Biochem. Vol. 7 (in press).
7. Henderson, R., Capaldi, R.A. and Leigh, J.S. (1977) J. Mol. Biol. 112, 631-639.
8. Fuller, S.D., Capaldi, R.A. and Henderson, R. (1979) J. Mol. Biol. 134, 305-312.
9. Deatherage, J.F., Henderson, R. and Capaldi, R.A. (1982) J. Mol. Biol. 158, 487-495.
10. Yonetani, T. (1961) J. Biol. Chem. 236, 1680-1688.
11. Darley Usmar, V.M., Alizai, N., Al-Ayash, A.I., Jones, G., Sharpe, A. and Wilson, M.T. (1981) Comp. Biochem. Physiol. 68, 445-458.
12. Malatesta, F., Georgevich, G. and Capaldi, R.A. (1983) in: Structure and Function of Membrane Proteins (Quagliarello, E. and Palmieri, F., eds.) Vol. 6, pp. 223-235, Elsevier, Amsterdam.
13. Sarti, P., Colosimo, A., Brunori, M., Wilson, M.T. and Antonini, E. (1983) Biochem. J. 209, 81-89.
14. Wilson, M.T., Lalla Maharaj, W., Darley Usmar, V.M., Bonaventura, J., Bonaventura, C. and Brunori, M. (1980) J. Biol. Chem. 255, 2722-2729.

15. Love, B., Chan, S.H.P. and Stotz, E. (1970) J. Biol. Chem. 245, 6664, 6668.
16. Robinson, N.C. and Talbert, L. (1986) Biochemistry 25, 2328, 2335.
17. Georgevich, G., Darley Usmar, V.M., Malatesta, F. and Capaldi, R.A. (1983) Biochemistry 22, 1317-1322.

MYOSIN SUBFRAGMENT 1 CATALYZES THE POLYMERIZATION OF ACTIN AT LOW IONIC STRENGTH

E. Grazi

Istituto di Chimica Biologica, Università di Ferrara, Ferrara (Italy)

The interactions of heavy meromyosin (HMM) and of the myosin subfragment 1 (S1) with both G-actin and F-actin have been widely investigated. HMM was reported to accelerate actin polymerization at low ionic strength (1) and to induce in G-actin the EPR spectrum typical of the F-actin conformer, the effect being maximal at an HMM/G-actin ratio of 0.25. S1 in this respect was found to be much less effective, the EPR spectrum changes being maximal for an S1/G-actin ratio of 1 to 1 (2). The interaction of either S1 or HMM with F-actin was also studied by numerous physical techniques. By measuring the quasi elastic light scattering of F-actin solutions it was found that the flexibility of the filament was increased by HMM, the effect being maximal at a 1:6 molar ratio of HMM to actin. S1 did not affect actin flexibility (3). F-actin flexibility was found, on the contrary, to be decreased when the interaction of either HMM or S1 was studied by saturation transfer EPR. In these experiments, the addition of the myosin fragments were found to increase 4 times the correlation time of the spin label bound to actin (4). The interaction of F-actin with S1 and HMM was also studied by measuring the static and the transient anisotropy decays of a fluorescent chromophore bound to F-actin. The mobility of the chromophore was strongly decreased in the presence of Ca^{2+} ions and increased in the presence of Mg^{2+} ions; in both cases the phenomenon was maximal at a myosin fragment to actin ratio between 1/25 or 1/50 (5).

The effect of HMM was found to be maximal at HMM/actin ratios between 1:4 and 1:10 also when either the fluorescence of ethenoadenosine diphosphate bound to actin (6) or the pulse fluorimetry in polarized light of a fluorescence probe attached to actin (7) or the amplitude of the EPR spectrum brought about in actin by the magnetic interaction of a cysteine bound spin label with the tightly bound manganous ion (8) were studied.

The common conclusion of these studies was to suggest that the myosin head, by interaction with the actin protomer, induced a conformational change propagating to a number of neighboring protomers of the same actin filament. Curiously, to explain the effect of substoicheiometric concentrations of S1 and HMM on F-actin, the alternative possibility was not taken into account: namely that the same S1 molecule could induce the conformational change of more than one protomer by interacting in rapid succession with many protomers of the same actin filament. Recently it was found that in 0.1 M KCl, pH 7.0 and 20°C, the half-life of the acto-S1 complex is about 3 seconds (9). We can propose, therefore, that any S1-induced conformational change of F-actin with a half-life longer than 3 seconds, will show saturation at S1 concentrations lower than those of the actin protomers and thus will simulate the propagation of the effect through the F-actin filament.

Evidence that the above mechanism may indeed operate came from our studies on the interaction, at 23°C, of 0.49 μM S1 and 12 μM ATP-G-actin in a medium containing 2 mM KCl, 44 μM $CaCl_2$, 44 μM ATP (initial concentration) and 2mM tris-HCl buffer, pH 7.5. After the addition of S1 to the solution, viscosity was found to increase, reaching the specific viscosity of 0.26 (final value) after 60 minutes of incubation. By comparing the specific viscosity of the S1-actin mixture with the specific viscosity of a 12 μM actin solution, fully polymerized by 2 mM $MgCl_2$, and assuming the same intrinsic viscosity for the two polymeric species, it was found that, for each S1 molecule in the solution, 12.5 molecules of actin had been organized in a filamentous structure. Electron microscopic observation showed that these structures were similar to F-actin.

The possibility was considered that the organization of actin filaments by S1 could proceed through the initial formation of a 1:1 S1/actin complex, followed by the recruitment of actin monomers and by the subsequent change of their conformation:

$$\mathrm{S1} + \mathrm{G} \longrightarrow \mathrm{S1\text{-}G^*}$$

$$\mathrm{S1\text{-}G^*} + n\ \mathrm{G} \longrightarrow \mathrm{S1\text{-}(G^*)_{n+1}}$$

where "G" is ATP-G-actin and "G*" is the actin conformer of filamentous actin.

For the reasons detailed below, such a long-range propagation of a conformational change is unlikely. It is known that, by cross-linking actin with p-NN'-phenylenebismaleimide, oligomers ranging in size from the dimer to the decamer are formed (10). Electron microscopic observation, however, shows that the cross-linked material retains the filamentous structure even at very low ionic strength (2 mM tris-HCl buffer, pH 7.5). Furthermore, it is found that a variable number of native monomers are trapped between the oligomers (the ratio of the number concentration of the monomers to the number concentration of the oligomers varies from 1.2 to 2.5) (11). We thus conclude that, at low ionic strength, the conformational change induced by the "sticky" ends of an oligomer propagates through 2.5 monomers, i.e. only a little further than through the adjacent monomer, if the double stranded nature of the filament is taken into consideration. The capability of S1 to organize in a filamentous structure 12.5 molecules of G-actin cannot be explained by a model in which the conformational change, induced by S1 on a first actin molecule, propagates through the growing actin filament. It can best be explained by the model:

$$\mathrm{S1} + \mathrm{G} \longrightarrow \mathrm{S1^*\text{-}G^*} \longrightarrow \mathrm{S1^*} + \mathrm{G^*}$$

$$n\ \mathrm{G^*} \longrightarrow \mathrm{(G^*)_n} \longrightarrow n\ \mathrm{G}$$

which only requires that the rate of the spontaneous conversion of G* into G is slower than the rates of the S1* into S1 and n G* into $(G^*)_n$ conversions.

Acknowledgment: This work was supported by a grant from the Ministero della Pubblica Istruzione.

REFERENCES

1. Yagi, K., Mose, R., Sakakibara, I. and Asai, H. (1965) J. Biol. Chem. 240, 2448-2454.
2. Cooke, R. and Morales, F.M. (1971) J. Mol. Biol. 60, 249-261.
3. Fujime, S. and Ishiwata, S. (1971) J. Mol. Biol. 62, 251-265.
4. Thomas, D.D., Seidel, J.C. and Gergely, Y. (1979) J. Mol. Biol. 132, 257-273.
5. Miki, M., Wahl, P. and Auchet, J.C. (1982) Biochemistry 21, 3661-3665.
6. Miki, M., Kouyama, T. and Mihashi, K. (1976) FEBS Letters 66, 98-101.
7. Tawada, K., Wahl, P. and Auchet, J.C. (1978) Eur. J. Biochem. 88, 411-419.
8. Lo Scalzo, J., Reid, G.H. and Weber, A. (1975) Proc. Nat. Acad. Sci. U.S.A. 72, 3412-3415.
9. Criddle, A.H., Geeves, M.A. and Jeffries, T. (1985) Biochem. J. 232, 343-349.
10. Knight, P. and Offer, G. (1978) Biochem. J. 175, 1023-1032.
11. Grazi, E. and Lanzara, V. unpublished results.

STRUCTURAL EFFECTS IN THE SELF-ASSOCIATION OF γ II-CRYSTALLIN

P. Stiuso, D. Pulcini, A. La Pegna, C. Roscigno, R. Ragone and G. Colonna
Istituto di Chimica e Chimica Biologica
I Facoltà di Medicina
Università di Napoli

The eye lens is a proteic glass, the properties of which depend on the supramolecular organization of lens specific proteins, the α-, ß- and γ-crystallins (1). In order to understand this proteic system, it is necessary first to describe the intermolecular interactions among crystallins and then consider the behavior of this complex framework in normal and/or pathological states, e.g. cataracts (1). Such interactions are of importance in driving the steric arrangement of the proteins inside the lens.

Recent results (2) on very concentrated solutions of calf γ-crystallin (up to 67 mg/ml) have shown that this mixture of different homologous gene products has a tendency to self-associate in the absence of other lens proteins.

In this communication we will investigate whether such interactions occur with the purified γII-crystallin.

MATERIALS AND METHODS

The eyes of 4-6 month-old calves were obtained from a local slaughterhouse. Completely clear and non-pathological specimens were selected. γII-crystallin was purified according to Bjork (3). Care was also taken to perform the experiments in the dark in order to avoid undesirable photochemical reactions. Protein concentration was determined using an absorption coefficient, $A_{1cm}^{1\%}$, of 21 (3). Absorption spectra were measured by means of a Cary 219 spectrophotometer and circular dichroism spectra by a Jobin-Yvon Mark III dichrograph. Protein fluorescence was obtained by excitation at 295 nm with a Perkin-Elmer 44 B spectrofluorometer.

RESULTS AND DISCUSSION

The effect of pH on the molecular stability of γII-crystallin was evaluated by circular dichroism in the far UV region and by tryptophan emission fluorescence. The protein resulted to be very stable. No shift in the absorption maximum and/or changes in the CD spectra reflecting structural modifications were observed.

Figure 1 shows concentration difference spectra (near UV CD and near UV absorption) measured for γII-crystallin in Tris buffer at neutral pH and low ionic strength. The UV concentration difference spectrum is reported in the lower part of the figure. By comparing the absorption spectrum of native II-crystallin in water at a concentration of 0.5 mg/ml with that at 2 mg/ml a characteristic difference was observed with bands centered at 284 and 278 nm. Modifications of this kind are generally attributed to perturbations in the environment of the tyrosine residues (4). The upper part of the same figure also shows the near UV CD concentration spectrum of γII-crystallin in the 260-300 nm range under the same experimental conditions. The difference bands in the 270-300 nm range must originate from the composite effect of the tyrosine and tryptophan transitions. In fact, it is well known that only tryptophan and tyrosine contribute to the circular dichroism in this spectral region (5). The CD spectrum is more complex, and an exact attribution of bands is beyond the scope of our work.

Tyrosyl as well as tryptophan residues seem to be involved. However, the results still suggest a

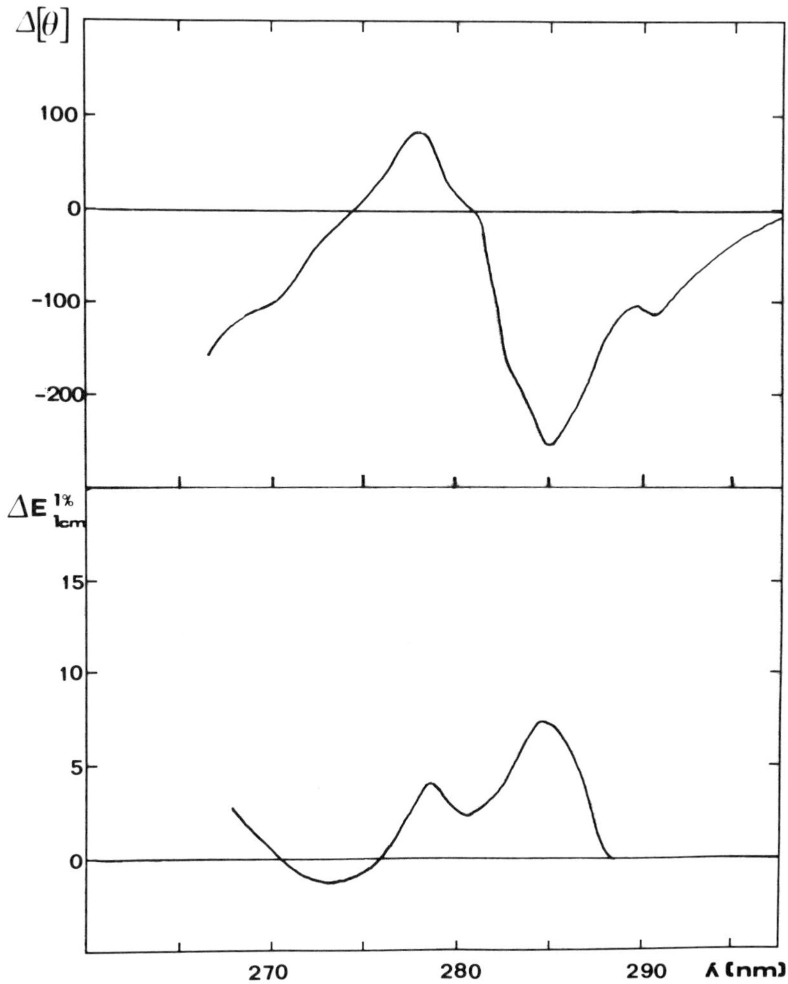

Figure 1. Concentration difference spectra of γII-crystallin.
Top: Near UV circular dichroism. 0.5 mg/ml solution in a 0.5 cm cell vs. a 2 mg/ml solution in a 2 cm cell. Buffer: 5×10^{-2} M Tris HCl, pH 7.5; temperature 20°C.
Bottom: UV absorption. Same experimental conditions as above.

structural perturbation of the environment of aromatic chromophores due to the effect of concentration.

A reasonable explanation for the observed spectral differences is that they reflect self-association. The phenomenon is strongly dependent upon the ionic strength. In fact, the interaction disappears by increasing the salt concentration up to 0.2 M KCl. This observation suggests an electrostatic basis for the interaction.

In conclusion, we found that pure γII-crystallin displays a greater tendency to interact at much lower concentrations than those found for the whole γ-fraction.

REFERENCES

1. Molecular and Cellular Biology of the Eye Lens (1981) (Bloemendal, H., ed.) J. Wiley, New York.
2. Siezen, R.J. and Owen, E.A. (1981) Biophys. Chem. 18, 181-194.
3. Bjork, I. (1964) Exp. Eye Res. 3, 254.
4. Wetlaufer, D. (1962) Advan. Protein Chem. 17, 303.
5. Chen, R.F., Edelhoch, H. and Steiner, R.F. (1969) in: Physical Principles and Techniques of Protein Chemistry (Leach, S.J., ed.) pp. 171-244, Academic Press, New York.

THE ROLE OF LYSINE-7 IN RIBONUCLEASE-A

Bruno Filippi, Gianfranco Borin and
Fernando Marchiori

Department of Organic Chemistry, Biopolymer Research Center, University of Padua
35131 PADUA, Italy

Bovine pancreatic ribonuclease A(RNase A) catalyzes the hydrolysis of RNA by a two-stage mechanism. The first step is a transphosphorylation to give an oligonucleotide terminating in a pyrimidine 2'-3'-cyclic phosphate. The second is the hydrolysis of the cyclic phosphate to give a terminal 3'-phosphate. At present it is not clear whether the invariant lysine-7 plays any part in the enzyme activity. Chemical modifications of the ε-ammonium group give derivatives with 15-30% activity, but substrate binding has been observed to be unaffected. On the other hand, the X-ray crystallography data, while confirming that Lys-7 is located relatively close to and with free access to the active site, strongly suggest that this residue is still too far away from the site to interact with the substrate (1).

In order to study the role of Lys-7 we prepared, by the conventional method of the peptide synthesis in solution, an S-peptide analogue (sequence 1-20 of RNase A), bearing the Lys-7/Dab replacement, in addition to our traditional Arg-10/Orn substitution. The side chain of the Dab residue is two-methylenic units shorter than that of Lys and, if an interaction occurs with the substrates, it is expected to alter the kinetic parameters.

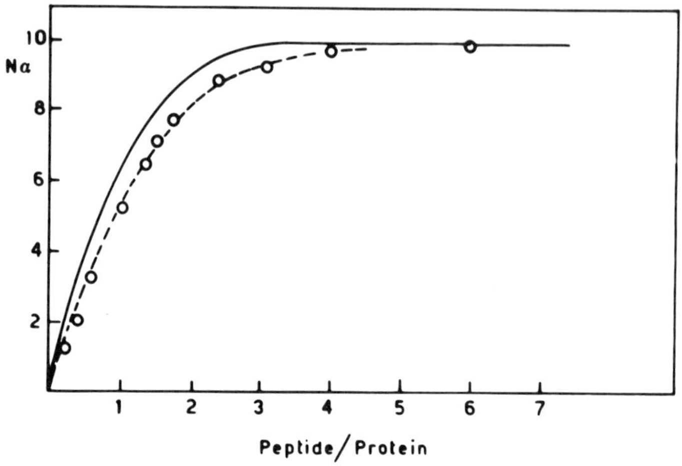

Fig. 1:CD titration of the S-protein by |Dab-7, Orn -10| (-o-) and by |Lys-7,Orn-10| (—) S-peptide analogues. The number of residues undergoing the coil to α-helix transition was calculated by the CD increment at 222 nm, using $|\theta|$mrw=-36000 for a residue in α-helical conformation.

The binding of the Dab-7 S-peptide to S-protein (sequence 21-124 of RNase A) was determined by Circular Dichroism (CD) measurements in the peptide absorption region. The comparison of the binding curves (Fig. 1) shows that the Lys-Dab replacement produces a small decrement of affinity to the S-protein. Since the Lys-7 ε-ammonium group does not give intermolecular interactions with the S-protein, the result reflects a peptide backbone conformational alteration (2); the Dab residue is known to have a lower α-helical propensity than that of Lys (3), and the coil to helix transition of the S-peptide constitutes the primary event in the binding to the S-protein.

The activation of the S-protein by the Dab-7 S-peptide, both in the transphosphorylation and in the hydrolytic step is less than that produced by Lys-7 (Fig. 2) and similar to that produced by the Ala-7 substitution (4). These results strongly evidentiate the need for the

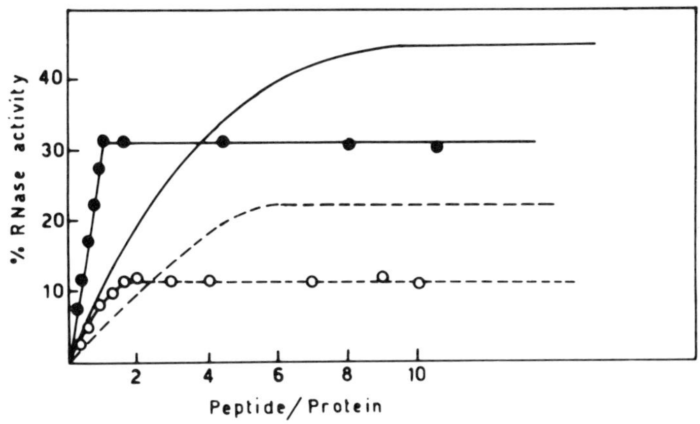

Fig. 2: Activation of the S-protein using RNA as substrate at pH 6.0(0.05 M dimehtylglutaric acid, 0.1 M NaOH taken to a total ionic strength of 0.2 with NaCl) by |Dab-7, Orn-10|-S peptide (-●-) and by |Lys-7, Orn 10|-S peptide (——). Activation of the S-protein using cCMP by |Dab-7, Orn-10| S-peptide (--o--) and by |Lys-7, Orn-10|-S-peptide (---).

ammonium group and the importance of its positioning for the best enzymic efficiency.

It is known that substrates stabilize the S-peptide/S-protein non covalent complex, and that the effect is greater for the high molecular weight RNA (1). The comparison of the dissociation constants (Kd) related to the |Lys-7, Orn-10| S-peptide/S-protein complex, obtained in the absence of substrate and in the presence of cCMP (Table I) may give a reasonable indication about the extent of this effect, since the data are obtained under comparable conditions. Therefore, the stabilization involving the Dab peptide is surprisingly high and is indicative of a very strong interaction with the S-protein, mediated by the substrates. The determination of the Michaelis constants and the comparison of the values rela-

TABLE I

		Dab	Lys
CD	Kd(M)	164×10^{-8}	120×10^{-8}
cCMP	Kd(M)	5.4×10^{-8}	84×10^{-8}
"	Km(M)	1.25×10^{-3}	1.12×10^{-3}
"	$K_2(s^{-1})$	0.38	0.68

ted to the Dab-7 and Lys-7 peptides (Table I) indicate that the Dab residue influences the break-down of the activated complex into the product, while it does not affect the stability Michaelis complex. The decreased enzymic efficiency depending upon the Lys-7/Dab substitution and the associated increment of binding suggests that in some step of the enzymic process following the formation of the activated complex, the γ-ammonium group is close enough to the substrate to form a strong hydrogen bond or an ionic interaction, by this way reducing the rate of break down into the products. This possibility was evidentiated by electronic structure theory calculations concerning Lys-41, for which a role similar to that of Lys-7 was proposed (5).

The unability of the X-ray data to evidentiate a role for Lys-7 depends upon the use of unreactive substrate analogues: the resulting enzyme/substrate complex is related to some early step of the enzymatic process, in which Lys-7 is not involved.

References

1. Richards, F.M. and Wyckoff, H.W. (1971) The Enzymes, 3rd Ed., 4, 647-806.
2. Komoriya, A. and Chaiken, I.M. (1982), J. Biol. Chem. 257, 2599-2604.
3. Grourke, M.T. and Gibbs, J.H. (1971) Biopolymers 10, 795-808.
4. Marchiori, F., Borin, G. and Moroder, L. (1974) Int. J. Pep. Prot. Res. 6, 419-434.
5. Deakyne, C.A. and Allen, L.C. (1979) J. Am. Chem. Soc. 101, 3951-3959.

SECTION II

INTERACTIONS OF NUCLEIC ACIDS WITH PROTEINS AND DRUGS

TRANSITION OF CHROMATIN FROM THE "10 nm" LOWER ORDER STRUCTURE, TO THE "30 nm" HIGHER ORDER STRUCTURE, AS FOLLOWED BY SMALL ANGLE X-RAY SCATTERING

K.O. Greulich, E. Wachtel, J. Ausio, D. Seger and H. Eisenberg

Polymer Department, The Weizmann Institute of Science, Rehovot 76100, Israel

INTRODUCTION

Double helical Watson-Crick B DNA is a stiff macromolecule which, in accordance with its basic structure (1), should proceed in free space in strictly linear fashion. Though abnormally migrating DNA has recently been discovered (2), and naturally curved DNA can presently be considered a highly fashionable topic of investigation (3), it must be considered an important exception, rather than the general rule. Still, even unrestricted in dilute solution, DNA is less rigid than could be gathered from its multiply fixed hydrogen bonded structure, and bends considerably as a result of disordering due to thermal Brownian motion. To describe the coiling behavior of DNA quantitatively the persistent chain or wormlike coil of Kratky and Porod (4) has been introduced. In this description the persistence length a equals about 40 nm in 0.2M NaCl (5), corresponding to a curvature radius of this order. Short DNA chains are almost rodlike in solution. Nucleosome length DNA (145 base pairs bp), comprising about one persistence length, is about 90% extended in solution, yet DNA chains comprising about 40 persistance lengths (6648 bp of the linearized plasmid ColE1 DNA for instance (6)) behave almost like randomly coiling gaussian chains (5). In interaction with histone proteins the DNA radius of curvature is much reduced and nucleosome core length DNA, for instance, achieves 1.75 superhelical turns, with a curvature radius of about 4.5 nm, in forming the nucleosome core particle by coiling around the histone octamer core (7).

In contrast to active globular proteins generated by folding amino acid chains comprising a few hundred amino acids and spanning a few nm only in space, DNA molecules of the higher organisms consist of linear arrays of the order of 48 to 240 million bp for the human chromosome, for instance, extending in length to about 1.6 to 8.2 cm (8), whereas the lateral dimensions are only about 2.5 nm. A structure of this kind cannot be maintained intact in unfolded form nor could it fit into the dimensions of a biological cell. The DNA molecule must thus be packaged in chromosomes in the form of chromatin to fit the dimensions of the cellular nucleus and yet this packaging must not be too tight, so as to prevent partial or local unfolding of chromatin for readout of the genetic message in either transcription or replication. We have already mentioned the first step in DNA compaction, i.e. complexing with the core histone proteins to form the nucleosome core particle, identified as the product of limited digestion of chromatin with micrococcal nuclease. In undigested, or only very mildly digested chromatin, the linker histones H1 and H5 "lock" 166 bp of DNA into two superhelical turns and futhermore participate in the folding and unfolding of chromatin into higher order or lower order structures, leading from less active, or more highly packaged, to more active, or less packaged forms (9). The structure of the histones comprises globular cores and randomly coiling positively charged extremities subject to post-synthetic modifications, such as acetylation or deacetylation of lysines and arginines, presumably affecting the above mentioned transitions.

The nucleosomes are spaced along the chromatin thread by so-called linker DNA stretches comprising from only a few up to a hundred bp, depending on the organism. Moreover, the distribution of linker lengths in a specific organism is variable rather than regular. *In vitro* the transition from the lower order LOS, to the higher order HOS structures is modelled by increasing NaCl concentration from a low value, about 1mM to about 80 mM; the transition may also be effected by small amounts of divalent ions such as Mg^{++}, for instance. At slightly higher concentrations of salt a major fraction of the chromatin aggregates and precipitates whereas at salt concentrations above 300 mM NaCl, first non-histone proteins (HMG proteins, for instance) and eventually the linker and then the core histones are displaced. So far it has not been established that once the linker histones are removed, they can be replaced in the native arrangement. Finch and Klug (10) introduced the concept that the transition from the LOS to the HOS proceeds from a nominally 10 nm diameter filament to a nominally 30 nm diameter single start contact helix (solenoid) with an 11 nm pitch and 5-7 nucleosomes per helix turn. However, even higher levels of coiling are necessary to achieve sufficient compaction.

The structure of chromatin of higher organisms can be studied in vivo inside intact nuclei, or the chromosome may be gently broken by mild DNAse action followed by isolation of chromatin fragments in solution. Typical chromatin fragments may comprise between 20 to 50 nucleosomes, though both shorter and longer fragments have been investigated. Bulk chromatin implies a random set of the total genome fragments.

Supporting evidence has been brought in favor of the solenoid model and evidence and arguments have been advanced in favor of alternate models such as the twisted-ribbon, the superbead, and a class of models in which the linker DNA, rather than proceeding along some predominantly circumferential path, crosses through the center of the cylinder, resulting in the absence of a central hole and a non-sequential arrangement of nucleosomes on the circumference (11). The still controversial status of the precise nature of the lower and higher order structures is well-illustrated in a number of very recent publications embodying unresolved conflicts both with respect to primary data and to interpretations. The current situation is reviewed in comments of Felsenfeld and McGhee (12).

Widom and Klug (13) have prepared gels of chicken erythrocyte chromatin fragments by concentrating chromatin solutions in the presence of polyvalent ions, in which chromatin is known to both fold and aggregate. They found that these samples were partially oriented in the X-ray capillaries and both meridional and equatorial reflections could be observed. These fiber patterns permitted Widom and Klug (13) to assign the orientation of the nucleosome in the HOS, supporting the Finch and Klug (10) solenoid model. In an additional contribution Widom et al (14) presented electron microscopy evidence to show that sea urchin sperm chromatin (repeat length about 240 bp of DNA) also forms 30 nm fibers, i.e. having the same diameter as the chicken erythrocyte fibers. They therefore concluded that the structure is invariant with respect to linker length.

Williams et al. (15), from a study of nuclei and unoriented chromatin solutions of seven different tissues under folding conditions, using electron microscopy and X-ray scattering, conclude that chromatin fibers are left handed, crossed linker double helices with diameter and mass per unit length that depend on linker length. The linker DNA criss-crossing the fiber section perpendicular to the fiber axis provides a natural explanation for the strong dependence they observe in the electron microscope of fiber diameter on linker length. Yet this is in contradiction to the observation of Widom et al. (14) that the fiber diameter (from two carefully studied cases) is independent of linker length.

Perez-Grau et al. (16) and Bordas et al. (17,18) have studied the organization of chicken erythrocyte chromatin and its condensation mechanism from the lower to the higher order structure by small angle X-ray scattering using synchrotron radiation. Their study is of particular interest in our work as, in contradistinction to the other authors, they dwell at length on the nature of the LOS. They observe, at salt concentrations below 30 mM NaCl, scattering patterns which, in addition to features already described by other authors (i.e. peaks due to the internal structure of the nucleosome and a linear region in plots of log s I(s) vs s^2 between s ~ 0.05 nm^{-1} and s ~ 0.1 nm^{-1}), exhibit a pronounced peak at s ~ 0.05 nm^{-1} in plots of log s.I(s) vs s, a minimum near 0.025 nm^{-1}, and a featureless increase at lower s values. (The scattering vector $s=(2/\lambda)\sin\vartheta$, where λ is the wavelength, 2ϑ is the scattering angle and I(s) is the scattering intensity). They interpret the peak at 20 nm as being due to periodic variations in electron density along the axis of the chromatin fiber, i.e. interference between individual nucleosomes separated by a distance of ~ 20 nm and arranged in a helix of pitch 33 nm, with 2.75-3.25 nucleosomes per turn. The 20 nm peak disappears, in the case investigated here of the chicken erythrocyte chromatin, with chromatin condensation, though Williams et al. (15) have (cf. above) associated a 20-25 nm peak for condensed chromatin from seven different tissues with the lateral dimension (diameter) of the chromatin fiber. Semi logarithmic plots of X-ray and neutron scattering from condensed chromatin always show two linear regions. The most condensed state of the chromatin fiber solution is, according to Bordas et al. (17,18) best represented by a crossed linked single start helical structure with about 2.56 nucleosomes per turn, a pitch of about 3 nm and a helical diameter of about 27 nm. This structure is in conflict with both the solenoid supported by Widom and Klug (13) and the double helical crossed linker structure of Williams et al. (15). As both (15) and (17,18) X-ray experiments have been undertaken with unoriented samples, it is not possible to resolve their differences in interpretation since in this case one is not able to distinguish unequivocally between meridional (along the fiber axis) and equatorial (perpendicular to the fiber axis) reflections.

The measurement of hydrodynamic parameters such as radii of gyration R_g , and frictional coefficients of fractionated chromatin samples can provide additional and independent constraints on models for uncondensed and condensed chromatin structure. In a recent detailed light scattering and hydrodynamic study (19) we followed the transition of defined chromatin fragments from chicken erythrocytes from the LOS to the HOS, in dilute, non-aggregating solutions. Properties determined (R_g from light scattering intensity, and frictional coefficients from quasi-elastic light scattering and sedimentation) characterize the overall size and shape of the chromatin particles.

In agreement with other workers we found (19) that at low chromatin concentration (A_{260} ~ 1), compaction of the LOS, as followed by a decrease in molecular friction, smoothly and steadily proceeds with increasing NaCl concentration above 1 mM (the process appears to be essentially complete around 75 mM NaCl) or with increasing $MgCl_2$ concentration, 0.3 mM $MgCl_2$ being roughly equivalent to 75 mM NaCl, at this chromatin concentration. From the angular dependence of scattered light intensity of chromatin fragments, at various ionic conditions, lack of aggregation was established and values of R_g were determined. Assuming a rigid solenoidal HOS at high salt with a cross-sectional radius of gyration, R_{gc} , equal to 9.3 nm (20) and a pitch, P, of 11 nm (21), we calculated, at 75 mM NaCl, a value of 5.7 for n, the number of nucleosomes per turn in the solenoid, for three fractions with z-average number of nucleosomes per chromatin fragment, N_z , ranging between 22 and 53. The calculation is not sensitive to the exact value of R_{gc} but is heavily dependent on the *a priori* assumption of the solenoid model value used for P. Some difficulties of interpretation occurred (19) when attempting to calculate frictional properties of this HOS on a quantitative basis. It was tentatively concluded that the assumption of a rigid, helical structure for the HOS, fitting the parameters given above, may be an oversimplification and that the real structure may be more complex. Thus we considered the possibility, proposed earlier by Fulmer and Bloomfield (22) that under the conditions of our experiments both the pitch P and the radius of the HOS may be considerably larger, and the structure is not a contact helix; there could be an intermediate structure in solution, between the LOS and a truly folded HOS in the resting nuclei.

Though helical structures in which consecutive turns are not touching may provide a seemingly reasonable interpretation of hydrodynamic and scattering results, the difficulty in rationalizing such structures resides in the absence of a suitable long-range force to provide the required stabilization. The realization therefore that helical symmetry, at least in the organization of chicken erythrocyte chromatin, is neither strictly prescribed from other experimental approaches nor derivable from the solution studies approach in our own work, relieved us of a certain bias imposed by current convention regarding the interpretation of biological structures. Helical structures may, in particular, be difficult to establish in chicken erythrocyte chromatin, due to the non-uniform length of the DNA linker connecting the nucleosomes (23). We have recently shown (24) that it is conceptually feasible to identify the folding process with gradual compacting of a freely jointed filament or a wormlike chain, within the limits of all the experimental methods used. This representation predicts not only hydrodynamic parameters but also results of X-ray small angle scattering experiments.

As the ionic strength is raised, the statistical length of the oligomer, composed of N_Z nucleosomes, decreases. At 75 mM NaCl, the compacted model chains (N_Z =53) form structures which are, on the average, cylindrically shaped with mean diameter 30 nm and length 104 nm. Helical symmetry need not be invoked in the modelling of the folding process. Concerning the shape of the X-ray scattering profiles at various salt concentrations, it is possible in this way to rationalize two-slope cross-section plots, also reported by other workers. Though this description represents a satisfactory conceptual presentation of a wealth of experimental data, it by no means represents a definitive solution to an exceedingly difficult problem.

RESULTS AND DISCUSSION

Small angle X-ray scattering data were obtained from fractionated chicken erythrocyte chromatin fragments prepared by a variety of procedures and determined with both one and two dimensional position sensitive detector systems. Essentially similar scattering results were obtained at both low and high salt for samples which had previously been characterized as "chromatin" or "soluble chromatin" (19) specifically referring to precipitation of the former material around 75 mM NaCl; the latter material was completely soluble without loss of linker histone components up to 150 mM NaCl. Optimal mixtures of NaCl and $MgCl_2$ for maximal compaction and total solubility at the concentration of the X-ray scattering experiment, have been previously defined (25). The argument that chromatin unfolding at very low salt concentration and subsequent attempts at refolding by appropriately increasing NaCl or $MgCl_2$ concentration may not completely restore the native HOS, has led to a variation in preparative procedure in which removal of $MgCl_2$ is avoided at all times (26). Following a suggestion by Williams et al. (15) electrolyte concentration in some preparations was never lowered below 5 mM EDTA which, according to these authors, should lead to restoration of native HOS upon increase of salt concentration. Experiments have been performed over a range of chromatin concentrations (between 1 to 8 mg chromatin/ml) to establish absence of concentration effects in the angular scattering range studied. The more dilute solutions were more homogeneous with respect to length as the samples could be obtained from a single tube from the sucrose gradient fractionation, rather than from a series of adjoining tubes. Length uniformity of the chromatin samples in cross-section plot analysis is not a stringent requirement.

Raw scattering data were plotted as I(s) vs s at low and higher salt, for samples prepared in various ways, as documented (24). Well defined peaks or shoulders were not discernible in most of these plots over the range of s values studied; internal nucleosome structure was not considered as the data do not extend to high enough s values. Obvious interparticle interferences were also absent, in the concentration range and in the values of s covered.

The data were replotted in the form of cross-section plots log q I(q) vs q^2 ($q=2\pi s$). In the range of salt concentrations between 1 and 40 mM NaCl the cross-section plots disclose a single linear region with d=1/s extending to 40 nm. (We define the low salt limit as 1 mM NaCl since below that value additional transformations or polyelectrolyte effects may lead to complex behavior). At the lowest values of q, there was observed in some cases a region of positive slope. We cannot say for sure if this is due to the effects of finite sample length and smearing, or if it is a true shoulder derived from the scattering curve. We address this question in more detail below. With further increase in NaCl concentration beyond 40 mM and up to 75 mM for chromatin solutions (up to 150 mM NaCl for soluble chromatin), and for solvent containing mixtures of NaCl and $MgCl_2$ a second linear region with larger slope appeared, at low values of q, with an inflection at d between 27-30 nm. Similar observations on small angle X-ray scattering of chromatin solutions exhibiting two slopes with increasing salt concentration and an identically placed inflection have been reported by other workers (27-29) and have been interpreted as arising from a binary mixture of cylinders, the smaller one characteristic of a LOS and the larger one characteristic of an HOS.

Before proceeding we reconsidered the question of the maximum in cross-section plots around d=20 nm, close to the upper limit of our experimental resolution range. Reich (30), in his preliminary study of X-ray scattering from the chromatin system in our laboratory, reported a further region at low salt with negative slope below d=40 nm. Bordas et al. (17,18) emphasized the maximum at 20-25 nm at low salt (which according to them disappears with chromatin folding at higher salt); their experiments with synchrotron radiation reach values around d_{max} =50 nm allowing visualization of the inner linear slope; Recently (S.E. Gerchman and V. Ramakrishnan, unpublished) neutron scattering cross-section plots of data from chicken erythrocyte chromatin to d_{max} =60 nm yielded at low salt a maximum around 20 nm and a second negative slope below 40 nm. At 70 mM NaCl they find a two slope behavior.

There have been attempts to explain the 20 nm feature at low salt on the basis of helical symmetry and to derive helical parameters on the basis of the position of the single peak (17,18). However, several independent arguments can be brought against the model of a helical structure at low salt. We calculate (24) for the low salt structure proposed by Bordas et al. (17,18) a radius of gyration R_g =184 nm for an oligomer of the size of fraction 5 of Ausio et al. (19). The measured R_g for this fraction of 1 mM NaCl is only 102 nm, so the actual structure must be much more compact. A helical structure of the type proposed by Bordas et al. (17,18) for low salt should yield a relatively narrow distribution of rotational relaxational times in electrical dichroism experiments. McGhee et al. (31) observe a broad spectrum of relaxation times at low salt which sharpens to a single relaxation spectrum upon folding. This supports the idea that the low salt structure is randomly coiling and not ordered. Stripping the linker histones H1 and H5, which releases constraints at the nucleosome DNA entrance and exit and abolishes folding at increased salt (32) does not abolish the 20 nm feature present at low salt (17,18, Gerchman and Ramakrishnan, unpublished).

Considering the peak as due to pair correlations with or without higher symmetry, we note that the maximum separation between nucleosomes for stretched out linkers

46 bp in length as in chicken erythrocytes does not reach 20 nm (31). Furthermore, the linker length in this chromatin is not uniform; thus a well defined interference term would not be favored. On the other hand, it is possible that a preferred curvature or other finite length feature of a structure which is not necessarily regular, may give rise to the 20 nm peak. In this case the angular region succeeding it, out to $q \sim 1/R_{gc}$, is still available for cross-section analysis.

If we therefore proceed with the classical line of analysis (33), then the mass per unit length of unfolded chromatin samples may be calculated from cross-section plots at low salt. The results of the evaluation for the LOS, yield a value $m_\ell = 2 \times 10^4$ g $mole^{-1}$ nm^{-1} between 1 and 40 mM NaCl. Considering 268,000 g $mole^{-1}$ for the molar mass of chicken erythrocyte chromatin nucleosome repeat unit (19), this leads to an average of one nucleosome per 13.4 nm along the LOS fiber axis. The nominal 46 bp linker is therefore almost maximally extended (85%), assuming a B-type conformation. We have calculated R_{gc} (LOS) from cross-section plots between 1 and 150 mM NaCl, from the higher angle linear portion of the cross-section plots; R_{gc} (LOS) slightly decreases from 2.5 to 2.1 nm in this range of NaCl concentrations. These observations are consistent with a coins-on-a-string like structure characterized by side-by-side packing of nucleosomes along the fiber axis. Similar conclusions were drawn from neutron scattering data obtained under various conditions of contrast (20) and from the sign of the electric dichroism (31). Thus the R_{gc} of the filaments does not increase with increase in NaCl concentration, and we did not observe the appearance of local structures corresponding to a zig-zag arrangement of nucleosomes. Such a transition would increase the cross section radius of gyration by approximately a factor of two. Zig-zag structures have been observed in electron micrographs and they may derive from a subtle sample-substrate relationship when flexible and delicate materials are examined by electron microscopy. The structures have been documented in a number of laboratories (34,35) and have led to the definition of pathways to HOS discussed above.

From the inner linear portion in the cross-section plots above 50 mM NaCl we derived values of R_{gc} which are related to the HOS. These values increase from about 6 to 8 nm. No variation in preparation protocol affected the characteristics of either linear region significantly.

While no apparent change in X-ray scattering is observed for values of d between 5 and 35 nm and for salt concentrations between 1 and ~ 40 mM NaCl, the same material is, in our hands (as well as reported by others), undergoing a folding process under these experimental conditions, as monitored by increasing diffusion and sedimentation coefficients, and a decrease in R_g. As a matter of fact, these changes occur promptly upon the slightest rise in NaCl concentration above 1 mM. This finding has been rationalized in the past by postulating that with increasing NaCl concentration an intermediate structure comes into being, possibly a helical structure with large pitch (22). However, we have already noted that for the interpretation of HOS by a helical structure with non-touching helical turns a suitable stabilizing force field is lacking. The two cylinder construct was devised to rationalize the presence of two linear regions in cross-section plots; some degradation was assumed to be the cause (29). However, as shown above, no modification of preparative procedure affected our results, and we take this as a strong argument against such an interpretation. We therefore proceeded to try to interpret our data in terms of the freely jointed filament of wormlike coil models. These, as will be demonstrated below, display features of SAXS curves experimentally observed. In view of the limitations of the theories used we attach only semiquantitive significance to the results of the calculations.

Values of a, the persistence length, and values of A, the statistical length, were fitted for the wormlike coil and jointed filament models respectively, using the light scattering R_g values previously reported (19). The values of 2a and of A are similar and both decrease sharply with NaCl concentration increasing from 5 mM to 75 mM; a consistent fit was achieved for the three fractions investigated. Values of 2a were used to calculate diffusion coefficients D from frictional coefficients f, calculated by the theory of Yamakawa and Fujii (36) for wormlike coils without excluded volume. A value 11 nm was used for the hydrodynamic diameter of the coil element. The fit is reasonable considering the neglect of excluded volume in both the derivation of the persistence length from the R_g values as well as in the calculation of the frictional coefficients. If excluded volume were properly accounted for, the calculated value of D would be expected to decrease.

SAXS curves were next generated using the appropriate values of A and of L, the total chain contour length. These curves, calculated by the jointed filament model, correspond to 5, 37.5 and 75 mM NaCl and two interpolated concentrations, for L=710 nm and R=3.2 nm. Timasheff et al. (37) have pointed out that the effect of connecting cylinders into compact structures is to increase somewhat the apparent cross-section dimensions; we have chosen a value for R which gives the observed LOS cross-section radius of gyration. We have plotted the model curves to d_{max} =60 nm, that is, beyond the region (d_{max}=40 nm) accessible to us experimentally. Two-slope plots are obtained at both low and higher salts (large and smaller filament). The outer linear region gives R_{gc} which varies from 2.4-2.8 nm. This corresponds to the local structure which remains unchanged during the folding procedure in agreement with the experimental results. The slope of the inner linear region up to d=40 nm increases with increasing salt in a similar manner to that observed experimentally. However, at lower angles a somewhat different behavior is predicted. Between d=50 nm and d=60 nm, an inner linear region with slope corresponding to R_{gc} =8.7 nm is found for 5 mM NaCl, but a cross-section plot is not strictly applicable here. This is because modelling with the jointed filament model has shown the average low salt structure to be irregular, approximating neither a cylinder nor an ellipsoid. The intermediate salt (37.5 mM NaCl) profile between d=50 nm and d=60 nm gives R_g =10-11 nm, as does the high salt (75 mM NaCl). For the latter, this slope appears roughly constant to the lowest angles for which a cross-section may be calculated. In this case, the scattering curve of the folded chains has the angular dependence of that of a cylinder with diameter 30 nm and length 104 nm. This may be compared to measurements on electron micrographs of similar length oligomers. The vector distribution function calculated by the inverse transformation procedure of Glatter (38) shows a prolate structure of maximum dimension of 130 nm and R_g=32.4 nm. In agreement with the views of others (Ref. 31, for instance) we thus model the transition from a low salt irregular, coiled structure to a higher salt, folded, cylinder-like average structure, although we cannot comment on the internal arrangement of nucleosomes and linkers.

The scattering from the wormlike coil is smooth; that from the segmented filament does display weak inflections whose position vary with A. This effect is likely due to the segmented nature of the model. Kirste (39) has shown that a disordered structure, such as a wormlike coil, will show oscillations in its scattering curve, if a certain curvature in the coil is favored. The degree of preference of the curvature is determined by the degree of rotation which the filament experiences. In some cases, the radius of the preferred curvature and the amount of preference of the curvature can be estimated. However, although the coil may look superficially like a helix, one cannot define a pitch, and the position of the scattering peak does not necessarily bear the same relationship to the radius of curvature as would be true for a perfect helix. Thus as shown in studies on synthetic and natural polymers, SAXS curves of structures with preferred curvature may display features which had previously been considered to be associated with helical structures (39,40).

CONCLUSIONS

The determination of the structural arrangement of chromatin fibers in either the LOS and HOS, as well as the transition between them, is difficult because of the size of the particles, which considerably exceeds that of its components (the nucleosome core particles, for instance) the structure of which can be determined, not without considerable effort, by X-ray crystallography (7). Furthermore, the arrangement is labile and dynamic, capable of easily adapting to the packaging versus functional role mentioned in the introduction. Low resolution solution methods, such as light scattering and sedimentation yield radii of gyration and frictional coefficients which can only be related in non-unique fashion to a given model. Electron microscopy, an excellent tool for rigid, well-defined, stable structures, must be considered with care in view of the delicate nature of the chromatin structures. Thus, for instance, an intermediary zig-zag structure of nucleosomal arrangement, documented in electron microscopy, was not observed by the analysis of small angle X-ray scattering (24). We used small angle X-ray scattering to increase the resolution of the earlier solution studies as well as to limit the range of model structurs deducible from the solution studies. SAXS is undertaken at considerably higher concentration than light scattering and sedimentation and this was considered in the interpretation of the experimental data, both with respect to interparticle interactions as well as in characterizing the role of low-concentration ionic solution components. While the use of X-ray scattering in solution indeed considerably limits the range of models derivable from lower resolution methods, it cannot provide a uniquely defined high resolution solution, and in particular the _deus ex machina_ introduction of helical features, which are not easily documented, may lack justification.

We have attempted to interpret the experimentally observed folding of fractionated chromatin oligomers in dilute solution with a minimum of modelling assumptions. It was shown to be feasible to explain the totality of solution data by a simple method involving the transition of a freely jointed filament or a wormlike coil from a less to a more compact form. As the ionic strength increases, the filament segment length of the oligomer decreases. At small angles, as in the case of light scattering, the scattering is not sensitive to the local structural details of the filament, seeing only the overall dimension of the coil. At large angles, the scattering appears qualitatively like that of disconnected randomly oriented cylinders with cross-sectional dimensions and mass per unit length characteristic of the filament itself. With this model we hope we may extend current thinking about what is still an important, unsolved problem.

Acknowledgements. This study was supported by grants from the Israel-U.S. Binational Science Foundation, Jerusalem, Israel and from the Minerva Foundation, Munich, Germany. K.O.G. was supported by a grant from the Minerva Foundation, Munich, Germany.

REFERENCES

1. Watson, J.D. and Crick, F.H.C. (1953) Nature 171, 737-738.
2. Simpson, L., Simpson, A.M., Kidane, G., Livingston, L. and Spithill, T.W. (1980) Am. J. Trop. Med. Hyg. 29(5) Suppl., 1053-1063.
3. Eisenberg, H. (1986) TIBS 11, 350-351.
4. Kratky, O. and Porod, G. (1949) Rec. Trav. Chim. Pays-Bas 68, 1106-1122.
5. Kam, Z., Borochov, N. and Eisenberg, H. (1981) Biopolymers 20, 2671-2690.
6. Chan, P.T., Ohmori, H., Tomizawa, J. and Lebowitz, J. (1985) J. Biol. Chem. 260, 8925-8935.
7. Richmond, T.J., Finch, J.T., Rushton, B., Rhodes, D. and Klug, A. (1984) Nature (London) 311, 532-537.
8. Kornberg, A. (1980) DNA Replication, W.H. Freeman and Company, San Francisco, CA.
9. Bradbury, E.M., Maclean, N. and Matthews, H.R. (1981) DNA, Chromatin and Chromosomes. John Wiley and Sons, Inc., New York.
10. Finch, J.T. and Klug, A. (1976) Proc. Natl. Acad. Sci, USA 73, 1897-1901.
11. Staynov, D.Z. (1983) Int. J. Biol. Macromol. (, 3-9.
12. Felsenfeld, G. and McGhee, J.D. (1986) Cell 44, 375-377.
13. Widom, J. and Klug, A. (1985) Cell 43, 207-213.
14. Widom, J., Finch, J.T. and Thomas, J.O. (1985) EMBO J. 4, 3189-3194.
15. Williams, S.P., Athey, B.D., Muglia, L.J., Schappe, R.S., Gough, A.H. and Langmore, J.P. (1986) Biophys. J. 49, 233-248.
16. Perez-Grau, L., Bordas, T. and Koch, M.H.J. (1984) Nucl. Acids Res. 12, 2987-2996.
17. Bordas, J., Perez-Grau, L., Koch, M.H.J., Vega, M.C. and Nave, C. (1986) Eur. Biophys. J. 13, 157-174.
18. Bordas, J., Perez-Grau, L., Koch, M.H.J., Vega, M.C. and Nave C. (1986) Eur. Biophys. J. 13, 175-186.
19. Ausio, J., Borochov, N., Seger, D. and Eisenberg, H. (1984) J. Mol. Biol. 177, 373-398.
20. Suau, P., Bradbury, E.M. and Baldwin, J.P. (1979) Eur. J. Biochem. 97, 593-602.
21. Sperling, L. and Klug, A. (1977) J. Mol. Biol. 112, 253-263.

22. Fulmer, A.W. and Bloomfield, V.A. (1982) Biochemistry 21, 985-992.
23. Compton, J.L., Bellard, M. and Chambon, P. (1976) Proc. Natl. Acad. Sci. USA 73, 4382-4386.
24. Greulich, K.O., Wachtel, E., Ausio, J., Seger, D. and Eisenberg, H. (1987) J. Mol. Biol. in press.
25. Borochov, N., Ausio, J. and Eisenberg, H. (1984) Nucl. Acids Res. 12, 3089-3096.
26. Sanders, M.M. (1978) J. Cell. Biol. 79, 97-109.
27. Hollandt, H., Notbohm, H., Riedel, F. and Harbers, E. (1979) Nucl. Acids Res. 6, 2017, 2027.
28. Damaschun, H., Damaschun, G., Pospelov, V.A. and Vorob'ev, V.I. (1980) Molec. Biol. Rep. 6, 185-191.
29. Brust, R. and Harbers, E. (1981) Eur. J. Biochem. 117, 609-614.
30. Reich, M.H. (1982) Ph.D. thesis, The Weizmann Institute of Science, Rehovot, Israel.
31. McGhee, J.D., Rau, D.C., Charney, E. and Felsenfeld, G. (1980) Cell 22, 87-96.
32. Allan, J., Staynov, D.Z. and Gould, H. (1980) Proc. Natl. Acad. Sci. USA 77, 885-889.
33. Sperling, L. and Tardieu, A. (1976) FEBS Letters 64, 89-91.
34. Thoma, F., Koller, Th. and Klug, A. (1979) J. Cell. Biol. 83, 403-427.
35. Woodcock, C.L.F., Frado, L.L.Y. and Rattner, J.B. (1984) J. Cell. Biol. 99, 42-52.
36. Yamakawa, H. and Fujii, M. (1973) Macromolecules 6, 407-415.
37. Timasheff, S.N., Witz, J. and Luzzati, V. (1981) Biophys. J. 1, 525-537.
38. Glatter, O. (1979) J. App. Crystallogr. 12, 166-175.
39. Kirste, R.G. (1967) in: Small Angle X-Ray Scattering (Brumberger, H. ed.) Gordon and Breach, New York, pp. 33-61.
40. Kirste, R.G. and Oberthür, R.C. (1982) in: Small Angle X-Ray Scattering (Glatter, O. and Kratky, O., eds.) Academic Press, London, pp. 387-431.

TOPOLOGY OF DNA IN ITS INTERACTION WITH ACTINOMYCIN D AND WITH THE HISTONE OCTAMER

Maria Savino["°], Pasquale De Santis[+], Luisa Leoni[°], Carmela Palermo[°] and Antonio Palleschi[+]
["]Dipartimento di Genetica e Biologia molecolare, [+]Dipartimento di Chimica, [°]Centro per lo Studio degli Acidi Nucleici
Università di Roma "LA SAPIENZA" - Roma, Italy

Large experimental evidence has recently well established that DNA in biological systems is supercoiled and that its behaviour is governed by topology (1,2).

Supercoiling results when two continuous strands in a duplex DNA ring are wound a number of times each around the other. This number is called the linking number, Lk, and it is a topological invariant, since it cannot change unless the continuity of the strands is interrupted. The straight connection between the linking number and the double helix structure of DNA is expressed in the equation:

$$Lk = Tw \qquad (1)$$

which applies when the DNA molecule is imagined projected on a flat surface (relaxed) and where Tw (twisting) is the number of helical turns in the DNA duplex. If the molecule, as is the real case, is free in space, equation (1) becomes:

$$Lk = Tw + Wr \qquad (2)$$

where Wr (writhing) is represented by the number of DNA superhelical turns. Since Lk is a topological invariant, from eq. 2 it follows that the sum Tw + Wr must remain constant through all structural transformations of DNA, unless they cause the breaking of phosphodiester bonds, while the values of Tw and Wr can variate in concert. As a consequence, DNA interacting with proteins or small molecules which can influence writhing, (histones in chromatin) or twisting (intercalators, such as ethidium bromide, proflavine etc.) in isotopological transformations can perform large variations of secondary structure and/or superstructure. The concept of topological domain (namely a system in which every perturbation is transmitted elastically and integrated) requires these interactions, which are local (i.e. are restricted to a definite nucleotide sequence), to be transmitted along DNA. Thus their effects can result in conformational transitions of distant nucleotide sequences (3).

The mechanisms adopted by molecules capable of perturbing the DNA equilibrium of twisting and writhing are very different. We want here to illustrate specific interactions between DNA and a relatively small molecule, actinomycin D, and the interactions between DNA and the histone octamer. Both molecules strongly influence DNA supercoiling albeit by means fo a completely different mechanism.

MATERIALS AND METHODS

pBR322 plasmid was prepared from E. coli by standard procedures.

Actinomycin D (AMD) was a gift from Merck and Co. Inc. Solutions were prepared by direct weighing and dissolving in double distilled water; the concentrations were checked spectrophotometrically, assuming $E_{440} = 22{,}000\ M^{-1}\ cm^{-1}$.

Topoisomerase I was prepared from chicken erythrocyte nuclei and used in standard buffer conditions (4).

Restriction enzymes were purchased from New England Biolabs and used in standard buffer conditions.

Histone octamer was prepared from chicken erythrocyte nuclei as previously described (5).

AMD-pBR322 DNA complexes were prepared by addition of convenient amounts of drug to DNA dissolved in 10^{-2} M Tris,

10^{-3}M EDTA, pH 7.5 (TE buffer), followed by incubation in the dark for 30 min.

Reconstitution procedure

A constant amount of plasmid DNA was added at 0°C, while stirring, to varying amounts of histone octamer at a final concentration of 1M NaCl, the salt concentration was lowered gradually to 0.1M NaCl and finally the samples were dialyzed overnight against the buffer required for the M.E. visualization (6).

Psoralen cross-linking and electron microscopy

15 µl trimethylpsoralen were added to pBR322 DNA complexed with the histone octamer at the ratio of 0.5 in 10 mM triethanolamine pH 8, 10 mM NaCl on ice, the irradiation was carried out as reported in (6). The psoralen-cross linked DNA was denatured and spread according to Sogo et al. (7). The complexes at the histone/DNA ratio (w/w) of 0.6 were prepared by the diffusion technique as reported by Thoma et al. (8).

Two dimensional electrophoresis were performed according to Wang et al. (9).

TOPOLOGICAL ASPECTS OF INTERACTIONS BETWEEN CIRCULAR DNA AND ACTINOMYCIN D

Actinomycin D (AMD) is a rather small molecule, characterized by an aromatic planar phenoxazone system and two cyclopentapeptide lactone rings (Fig. 1), with specific therapeutic activity (10), generally attributed to interactions

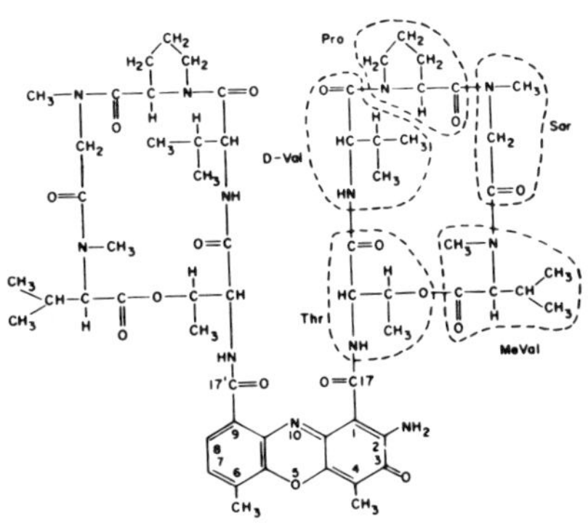

Fig. 1 - AMD molecular formula.

with DNA (11). Recently we have shown that DNA is capable to intercalate in nucleosomes and has a dramatic effect on their hydrodynamic behaviour (12). However, a molecular model of complex between DNA and AMD, supported by direct experimental evidence, is still lacking. Indirect informations derive from studies on model systems, namely complexes between AMD and a variety of oligonucleotides. Solution studies of these complexes indicate that the AMD planar chromophore intercalates between base pairs in the double helix, the preferred binding sequence being GC (13). More recently Takusagawa et al. (14) determined the structure of AMD and d(GpC) complex by crystallographic analysis. The structure revealed some features of interaction rather different from the classic intercalation: namely the AMD molecule intercalates its phenoxazone ring between two guanines belonging to two different double strand chains, so that the complex can be considered pseudointercalated. To obtain experimental evidence in solution on this structure, which could give rise to a sort of non covalent cross-linking of DNA, we have studied the binding of AMD to the supercoiled DNA of the plasmid pBR322 at very low AMD/DNA molar ratios using a topological approach. This consists in the freezing of the topological state of the DNA-complex with AMD, by the reaction of breaking and resealing of phosphodiester bonds catalyzed by the enzyme topoisomerase I. Fig. 2a shows that in absence of AMD the reaction produces fully relaxed DNA, whereas even at very low drug to polynucleotide molar ratios (e.g. 1/100) AMD is able to stabilize at least ten supercoils. For AMD/DNA ratios higher than 1/100, the resolution of agarose gel is too low for counting the number of supercoils. In this case we have used the band shift method (15), which takes advantage of the fact that a classic intercalator, namely ethidium bromide, by untwisting DNA of 26° for each molecule bonded, is capable of decreasing the DNA supercoils by a determined number. In Fig. 2b ΔLk evaluated by this method is reported as a function of the AMD/DNA input molar ratio (r). A peculiar trend can be seen: the values corresponding to the lower r values potentially could be originated by a hypothetical DNA untwisting of 90°

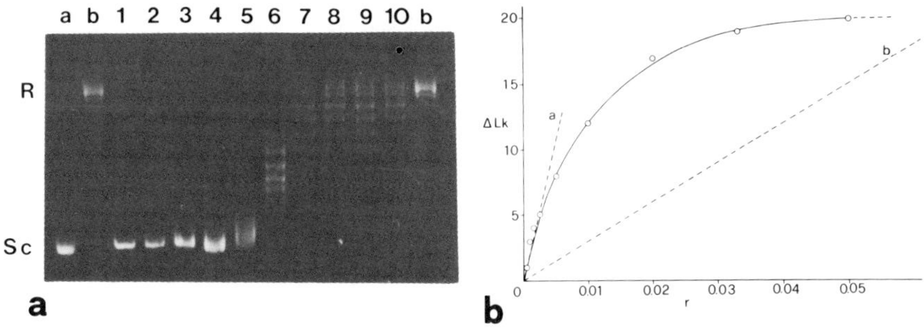

Fig. 2a Agarose gel electrophoresis of complexes between supercoiled pBR322 DNA and AMD at different drug/DNA input molar ratios (r), after topoisomerase I reaction and DNA purification.
a) DNA supercoiled b) r = 0; 1) r = 0.1; 2) r = 0.05; 3) r = 0.03; 4) r = 0.02; 5) r = 0.01; 6) r = 0.005; 7) r = 0.003; 8) and 9) r = 0.002; 10) r = 0.001.
Fig. 2b Variation of the linking number (Lk) of supercoiled pBR322 DNA in complexes with AMD as function of r. ΔLk was evaluated from results reported in Fig. 2a, except for higher r (see text). The dotted line (b), represents the expected theoretical trend, if only the intercalative binding is considered with an untwisting of 28°.

(see lane a), if the generally accepted intercalative binding is assumed. By increasing r,the curve of the experimental ΔLk becomes closer to the theoretical line corresponding to the untwisting of 28° (16). The two trends are coincident for r = 1/20.

Since an interaction between the enzyme topoisomerase I and AMD cannot be excluded we have carried out similar experiments on pBR322 DNA previously relaxed, by topoisomerase I. Fig. 3a shows that the features of agarose gel electrophoresis of relaxed DNA are almost equal to those of supercoiled DNA. Furthermore in Fig. 3b, the trend of ΔLk as a function of r is similar to that of Fig. 2b, even though the actual ΔLk values of the complexes are closer to the theoretical intercalation line.

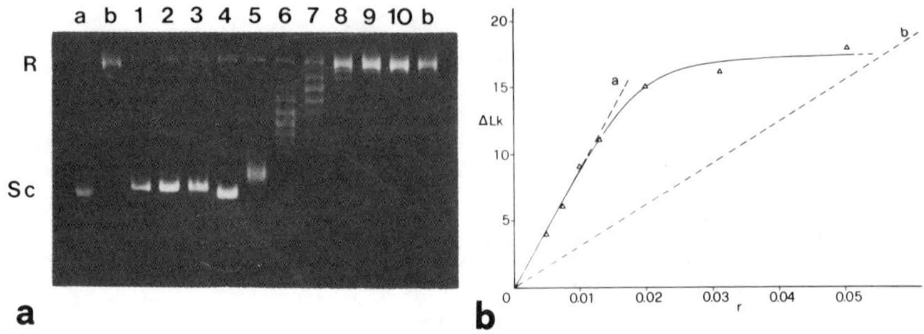

Fig. 3a Agarose gel electrophoresis of complexes between relaxed pBR322 DNA and AMD at different r, treated as in fig. 2a.
a) DNA supercoiled; b) DNA relaxed; 1) r = 0.1; 2) r = 0.05; 3) r = 0.03; 4) r = 0.02; 5) r = 0.01; 6) r = 0.005; 7) r = 0.003; 8) and 9) r = 0.002; 10) r = 0.001.
Fig. 3b Variation of the linking number (Lk) of relaxed pBR322 DNA in complexes with AMD as function of r. ΔLk was evaluated as reported in Fig. 2b.

These results suggest that at very low AMD/DNA binding ratios specific interactions different from intercalation underlie the effect of AMD on DNA topology. As reported in (14) AMD could bind to DNA by pseudo intercalation, connecting two different double strands. According to this model, the topological effect of AMD at very low r could derive from the stabilization of the crossings of supercoiled DNA. To test this hypothesis, we determined the handedness of relaxed DNA complexed with AMD, after treatment with topoisomerase I, by bidimensional electrophoretic analysis in the presence of ethidium bromide. Taking into account that the electrophoretic mobility of positive supercoiled DNA is increased by ethidium bromide, whereas that of negative supercoils is lessened, the results reported in Fig. 4 demonstrate that AMD stabilizes negative DNA supercoils also in the case of relaxed DNA, which exists as an equimolar equilibrium of negative and positive supercoils fluctuating between +2 and -2 (17).

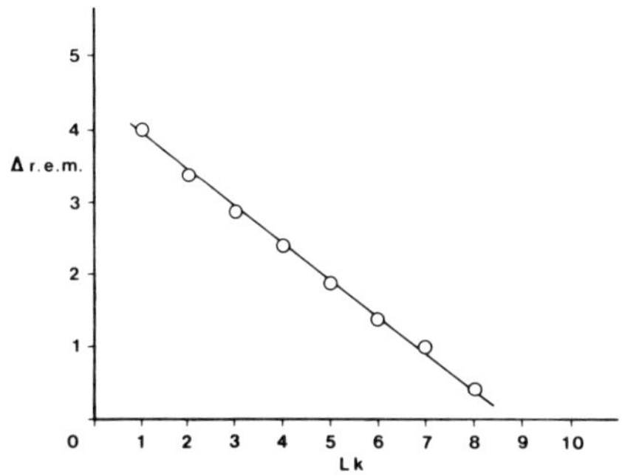

Fig. 4 Δ(rem) is the differential electrophoretic mobility of pBR322 topoisomers in the presence of ethidium bromide (c = 0.02 μg/ml) with respect to the nicked plasmid. The experiment was carried out on previously relaxed pBR322 DNA.

The topological analysis of the binding of AMD to negative supercoiled as well as to relaxed pBR322 DNA at very low binding ratios has revealed that this relatively small molecule can induce a dramatic change in the equilibrium between the twisting and writhing of circular DNA, far larger than that deriving from the DNA untwisting of 28° generally considered until now. Furthermore, the negative sign of Tw or Wr demonstrates that the effect relies on hypothetical 90° untwisting and/or stabilization of DNA negative supercoils.

Taking into account these results as well as those reported by Takusagawa et al. (14), we propose the model illustrated in Fig. 5; the basic feature of this model is the stabilization of the left-handed crossing of negative DNA supercoil. This is in agreement with the chirality of the left handed crossing of the DNA and with the fact that, as we have previously reported (18), AMD is a mixture of two opposite atropisomers that can be resolved by the stereospecificity of the interaction with DNA (19,20). The

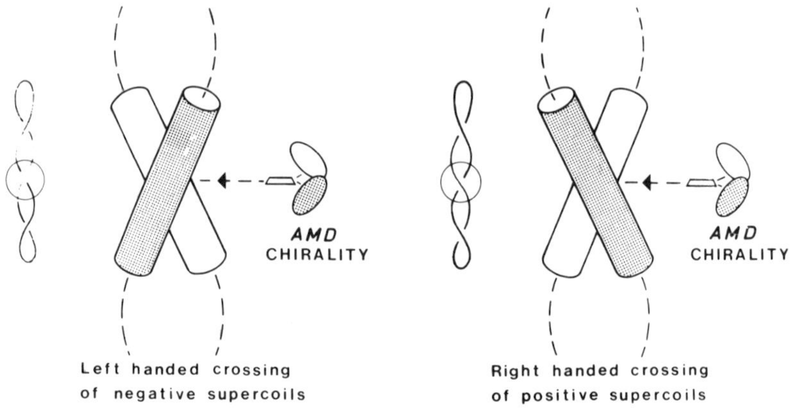

Fig. 5 Schematic drawing of the chiral discrimination between (a) left handed crossing of negative DNA supercoils and (b) right handed crossing of positive DNA supercoils by AMD.

freezing of DNA writhing at very low r requires AMD to be able of titrating specifically DNA supercoils at low binding ratios, in alternative to untwisting which prevails at higher binding ratios.

It is wortwhile to speculate that the first type of interaction could cause the strong inhibition of transcription by AMD at low r, which is known to start with the untwisting of promoter.

NUCLEOSOMES POSITIONING ANALYSIS BY ELECTRON MICROSCOPY VISUALIZATION AFTER PSORALEN CROSSLINKING AND DNA BENDABILITY

A few years ago, Zachau and Igo Kemenes (21) reported in the minireview entitled "Face to phase with nucleosomes" these remarks: "... Recent reports more or less agree that, in the system investigated, there exists a preference of the nucleosome for certain position. The extent of this preference, its structural reasons and its functional significance, however, are still far from being understood". In 1981 the main difficulty for imaging the structural reasons for the preferential positioning of histone octamers along the DNA sequence, derived from the concept of DNA structure that was generally accepted. DNA was imagined in a rather uniform B conformation, and thus inert in the recognition mechanisms by histone octamers. In fact, the author's concluding remarks speculate that determinants of nucleosome alignment may reside in the histone octamer itself and that "they may be accentuated by histone modification and the occurrence of histone variants". The general view of the problem has now been dramatically modified by the concepts of DNA bending and bendability (22).

In this paper we want to report the analysis of nucleosomes positioning in a non saturated system, in which the DNA structural determinants should be emphasized, using an experimental approach based on electron microscopy visualization after psoralen cross-linking. The results will be described, in relation with DNA bendability, on the basis of the analysis of nucleotide sequence fluctuations, which use as bendability matrix, dinucleotides differential deformations calculated from conformational analysis in our laboratory (23). Circular supercoiled pBR322 DNA was complexed with histone octamer at various histone/DNA ratios and nucleosome formation was tested by different assays (13).

Fig. 6a shows an electron micrograph obtained after fixation of reconstituted minichromosomes with glutaraldehyde. The chromatin characteristic pattern of beads on a ring is fairly evident. The positioning of nucleosomes along the nucleotide sequence was evaluated by trymethylpsoralene photoreaction of the reconstituted minichromosomes using methods reported in (13). In Fig. 6b the electron micrograph of pBR322 DNA, complexed at a ratio of 0.3 and treated with trimethylpsoralen is shown. Denatured bubbles are evident, whose length corresponds mainly to 140 bp (13); the bubbles appear clustered in some DNA regions and by the analysis of significative groups of molecules (at least 40), restricted with different enzymes, we have found that 75% of the molecules presents some significantly denatured regions

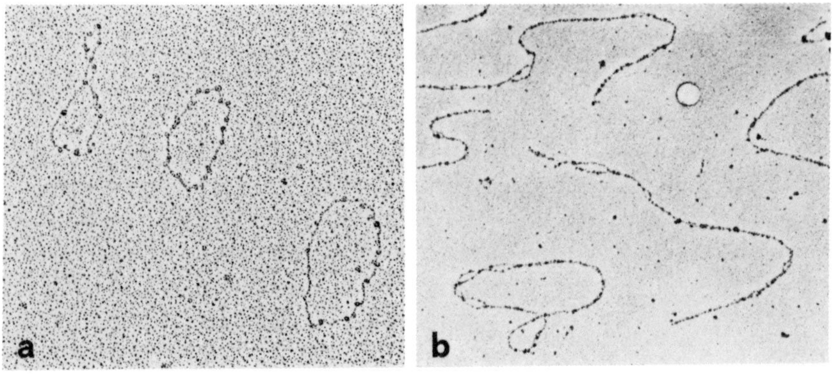

Fig. 6a Electron microscopy visualization of histone octamer and supercoiled pBR322 DNA, complexed at a ratio, Γ, (w/w) of 0.6.
Fig. 6b Electron micrographs of complexes between histone octamer and supercoiled pBR322 DNA at r = 0.3, after psoralen treatment and Pvu II linearization.

located at fixed distances from the ends (13,24).

ANALYSIS OF BASE SEQUENCE PERIODICAL FLUCTUATIONS

Regular sequence fluctuations characterized by the B DNA period were first observed by Trifonov (25). He calculated the autocorrelation functions of a variety of eukariotic and viral DNAs and showed that some dinucleotides have a clear tendency to be repeated periodically along the chromatin DNA sequences: the dinucleotides showing more often this feature were ApA and TpT. This was interpreted by Trifonov as a cause of DNA bending. Using a matrix of bendability derived from the analysis of chromatin DNA sequences, the author proposed a mapping procedure to locate nucleosomes along nucleotide sequences.

A different algorithm of locating the potential or actual bent of DNA was later proposed by Zurkin (26), based on the predicted distortions of the structures of PurPyr and PyrPur dinucleotide pairs according to the Calladine Dickerson rule (27). Recently the previsions based on the reported methods were tested by the experimental results of Crothers and coworkers (28,29) and Hagerman (30) using a very simple, but powerful electrophoretic method on natural as well as on synthetic DNAs. These authors, independently suggested the existence of cooperative effects since the simple consideration of the dinucleotide properties did not fully explain the experimental data. On the basis of conformational energy calculations of dinucleotide fragments of DNA, very recently De Santis et al. (23) identified that nucleotide bendability derives essentially from DNA roll, while the wedge deformation of base pairs is energetically unfavourable. These features of the DNA sequence can be to a first approximation, quantified in a bendability matrix, which assigns to the different 16 dinucleotides a complex number, which represents the bendability value and the phase which takes into account the direction of DNA deformation. This algorithm was tested firstly to predict the relative bends of synthetic periodical DNAs investigated very recently by Crothers et al. and Hagerman. The correlation is quite satisfactory and explains consistently the dramatically anomalous behaviour of such periodical DNAs (29,30).

Fig. 7a shows the bendability diagram of the circular DNA pBR322, 4360 bp long. The bendability was averaged over eighty base pairs and the relative phases were plotted only in the most significant regions where the bendability is higher than the sum of the average bendability plus the mean standard deviation. The bendable regions are easily localized along the sequence where the phases are nearly invariant, over more than 50 bp, as indicated by circles; these regions locate the preferential positioning of nucleosomes. Fig. 7b shows the frequency of bubbles along the sequence of pBR322 DNA obtained from electron microscopy of psoralen crosslinked reconstituted minichromosomes, which indicate, as discussed before, the frequency of nucleosomes along the DNA sequence. It is easily recognized that there is a fairly good correlation between the two patterns, implying that bendability governs nucleosome positioning on

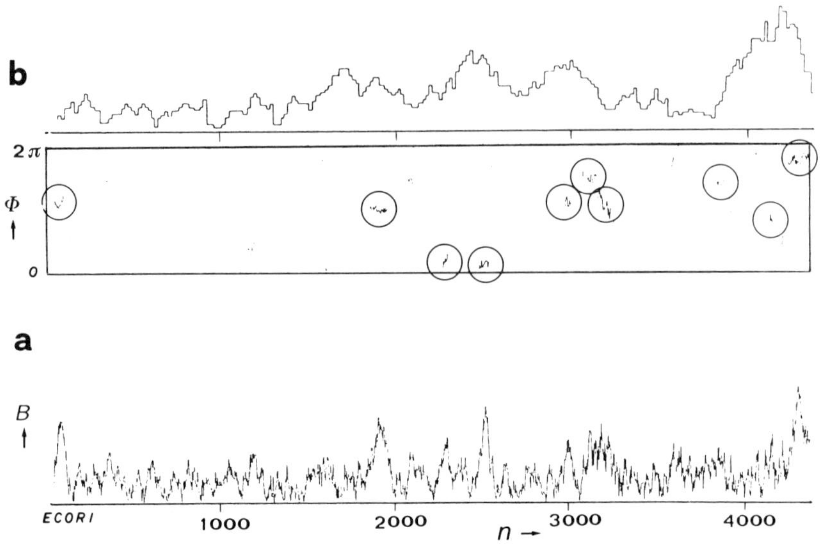

Fig. 7a Profile of bendability and relative phase of pBR322 the phase is reported only if the value is higher than the sum of average bendability and standard deviation.
Fig. 7b Histogram of locations of uncrosslinked bubbles on individual reconstituted minichromosomes that were treated with psoralen as reported in the text. The molecules were aligned from EcoRI site, also if different restriction enzymes cleavage was used from alignment (13).

DNA. Further, we have extended our analysis of bendability to SV40 DNA. This DNA is contained in Simian virus 40 (SV40) - infected cells in nucleoprotein structure with many similarities to cellular chromatin. Fig. 8 shows the bendability of SV40 DNA calculated with our matrix. The pattern is characterized by a more regular distribution of maxima along the sequence, which could be ascribed to the fact that SV40 DNA is organized in chromatin at variance with pBR322. Whether the features of the bendability profile imply that a number of nucleosomes has a more definite positioning is at the moment only a speculation. However, it is interesting that the nucleosome gap occurs at the beginning of

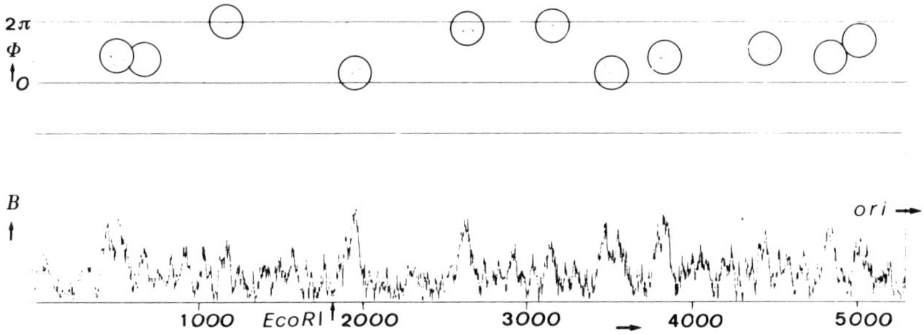

Fig. 8 Profile of bendability and relative phase of SV40 DNA.

the sequence. This region is adjacent to the viral origin of replication (ori site) and is hypersensitive to cleavage by endonucleases (31); it appears as a gap in the nucleosome pattern when visualized in the electron microscope (32).

CONCLUDING REMARKS

Until a few years ago DNA was thought of as a rather rigid molecule with a uniform, hardly bendable B structure and hence with an apathy in the recognition mechanism of proteins. Now the concept of DNA has been changed by the discovery of sequence dependent polymorphism which can be excited in connection with supercoiling density, and by the recognition of its potential or actual bendability as a property of the whole nucleotide sequence.

We have reported results showing that AMD, a rather small molecule, can influence strongly the equilibrium between writhing and twisting in a topological DNA domain at very low , binding biologically significant binding ratios.

Another example illustrates the topological aspects of the control process exerted by DNA itself on the winding on the histone octamer, which directs the DNA untwisting in a topological domain by blocking the writhing. The definite dependence of bendability on nucleotide sequence implies a control by DNA in the recognition process of chromatin proteins.

ACKNOWLEDGMENTS

Thanks are due to Dr. E. Caffarelli for electron microscopy visualizations and to Dr. G. Micheli for helpful advices on this technique. Thanks are also due to R. Gargamelli for help in drawings and photographs. This research has been partially supported by Grandi Progetti d'Ateneo dell'Università di Roma "La Sapienza" MPI 60% and by Fondazione Istituto Pasteur Cenci Bolognetti.

REFERENCES

1. Lilley, D.M.S. (1983) Nature, 305-276.
2. Fisher, L.M. (1984) Nature 307-686-687.
3. Crothers, D.M. and Fried M. (1983) Cold Spring Harbor Symp. Quant. Biol. 47, 263-269.
4. Bina-Stein, M., Vogel, T., Singer, D. and Singer, M.S. (1976) J. Biol. Chem. 251, 7363-7366.
5. Caffarelli, E., De Santis, P., Leoni, L., Savino, M. and Trotta, E. (1983) Biochim. Biophys. Acta 739, 235-243.
6. Caffarelli, E., Leoni, L., Sampaolese, B. and Savino, M. (1986) Eur. J. Biochem. 156, 335-342.
7. Sogo, J.M. , Ness, P.J., Widmer, R.M., Parish, R.W. and Koller, Th. (1984) J. Mol. Biol. 178, 897-919.
8. Thoma, F. and Koller, Th. (1977) Cell. 12, 101-107.
9. Wang, J.C., Peck, L.S. and Becherer, K. (1983) Cold Spring Harbor Symp. Quant. Biol. 47, 85-91.
10. Farber, S. (1966) J. Am. Med. Ass. 198, 826-836.
11. Gellert, M., Smith, C.E., Neville, D. and Felsenfeld, G. (1965) J. Mol. Biol. 11, 445-457.
12. Cacchione, S., Caneva, R. and Savino, M. (1986) Biochim. Biophys. Acta 867, 229-233.
13. Patel, D.J. (1974) Biochemistry 13, 2396-2402.
14. Takusagawa, F., Dabrow, M., Neidle, S., and Berman, H.M., (1982) Nature 296, 466-469.

15. Keller, W. (1975) Proc. Nat. Acad. Sci. USA 72, 4876-4880.
16. Waring, M.S. (1981) Ann. Rev. Biochem. 50, 159-192.
17. Pulleyblank, E.E., Shure, M., Tang, D., Vinograd, J., and Vosberg, H.P. (1975) Proc. Nat. Acad. Sci., USA 72, 4280-4284.
18. Ascoli, F., De Santis, P. and Savino M. (1970) Nature, 227, 1237-1239.
19. De Santis, P., Rizzo, R., Ughetto, G. (1972) Nat. New Biol. 237, 94-95.
20. Sobell, H.M. Sain S.C., Sakore, T.D., Nordman, C.E. (1971) Nat. New. Biol. 231, 200-205.
21. Zachau, H.G. and Igo Kemenes, T. (1981) Cell, 24, 597-598.
22. Lilley, D. (1986) Nature, 320, 487-488.
23. De Santis, P., Morosetti, S., Palleschi, A. and Savino M. Int. Symp. on Struct. and Dynam. in Biomolecules E. Clementi Ed. Plenum Press (1986) in press.
24. Caffarelli, E., De Santis, P., Leoni, L., Palleschi, A., and Savino, M. (to be published).
25. Trifonov, E.N. (1983) Cold Spring Harb. Symp. on Quant. Biol. 271-278.
26. Zurkin, V.B., Lysov, Y.P., and Ivanov, V.S. (1979) Nucleic Acids. Res. 6, 1081-1095.
27. Calladine, C.R. (1982) J. Mol. Biol. 161, 343-352.
28. Marini, J.C., Levene, S.D., Crothers, D.M. and Englund, P.T. (1982) Proc. Nat. Acad. Sci. U.S.A. 79, 7664-7668.
29. Koo, H.S., Wu, H.M. and Crothers, D.M. (1986) Nature, 320, 501-506.
30. Hagerman, P.J. (1986) Nature 321, 449-450.
31. Warshavsky, A.J., Sundin, O., and Bohn, M. (1979) Cell, 453-466.
32. Jakobovits, E.B., Bratosin, S. and Aloni, Y. (1980) Nature, 285, 263-265.

THERMODYNAMICS AND STEREOCHEMISTRY OF THE INTERACTION BETWEEN ANTHRAQUINONE DRUGS AND DNA

Manlio Palumbo[*], Giorgio Palu'[+] and Sebastiano Marciani Magno[§]
[*] Department of Organic Chemistry, [+] Institute of Microbiology, and [§] Department of Pharmaceutical Sciences, University of Padova, 35131 Padova, Italy

The majority of the drugs used today in cancer chemotherapy act as nucleic-acid synthesis inhibitors. The inhibitory process may occur either through direct interaction with the nucleic acid, giving a stable complex, or through interference with enzymes acting at the nucleic acid replication or transcription level (1,2). The anthracycline antibiotics Daunomycin and Adriamycin are among the most important derivatives for which the mechanism of action is believed to consist of a binding process to DNA. These compounds possess a tetrahydrotetracenedione chromophore, containing three coplanar hexaatomic rings. Numerous investigations have demonstrated the formation of an intercalation complex with DNA (2-5). At the molecular level, the aromatic ring system of the anthracycline slips into adjacent base pairs of the polynucleotide, whereas the cyclohexene moiety, connected to the amino-sugar is located in the minor groove of the double-helix (6,7). The complex is further stabilized by electrostatic interactions between the positively charged protonated amino group of the polymer chain.

In spite of their wide application in cancer chemotherapy, the efficacy of the above-mentioned drugs is limited by a low therapeutic index and the concomitant occurrence of undesired side-effects, such as bone marrow depression and cardiac toxicity (8). Many recent investigations aimed at developing new anthracycline analogues, preserving the ability of a strong interaction with DNA, but devoid of the above mentioned adverse effects. In particular the attention has been focused on substituted anthracenediones and naphtacenequinones, which maintain the essential structural features of the model drugs (9-16). Among them Mitoxantrone (MX), 1,4-dihydroxy-5,8-bis||2-(2-hydroxyethyl)aminoethyl|amino|-9,10-anthracenedione exhibited pronounced anticancer activity both *in vitro* and *in vivo* (17-19).

Due to the importance of an appropriate evaluation of the molecular events and structural requirements which lead to biological activity in anthraquinone derivatives, we recently synthesized a number of differently substituted 9,10-anthracenediones, whose chemical structure is reported in Fig. 1. The main features are the presence of 1 or 2 hydroxyl- (or ether) groups at positions 1 and 8, and of 1 or 2 positively charged side-chains at positions 1,2,4 and 5. Neutral substituents are possibly introduced at positions 3 and/or 4.

Our experimental approach includes a physico-chemical

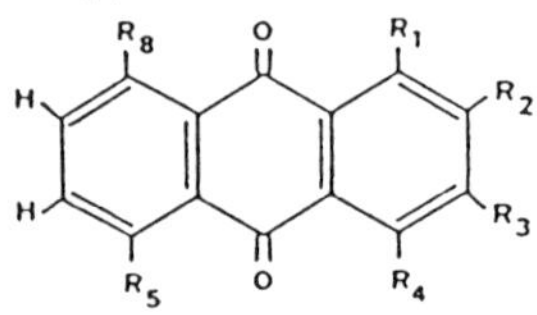

	I	II	III	IV	V	VI	VII	VIII	IX
R_1	OH	OH	Y	Y	OH	OH	OH	OH	OH
R_2	H	H	H	H	H	X	X	X	X
R_3	H	OCH_3	OCH_3	OCH_3	H	H	H	H	H
R_4	X	X	$NHCOCH_3$	NH_2	X	H	H	$NHCOCH_3$	X
R_5	H	H	H	H	H	H	X	H	H
R_8	H	H	H	H	OH	OH	OH	H	H

X = $-NHCO(CH_2)_2N(C_2H_5)_2 \cdot HCl$

Y = $-O(CH_2)_3N(C_2H_5)_2 \cdot HCl$

Figure 1. Chemical structure of anthraquinones I-IX.

characterization of the DNA-binding process, and an evaluation of the biological effects consequent to drug administration, which, together, lead to the proposal of new active structures. In this paper we will focus on the thermodynamic and stereochemical aspects of drug binding to DNA.

DNA-binding thermodynamics allows a quantitative evaluation of all parameters governing the interaction process. In particular the intrinsic binding constant, K_i, is related to the free energy change caused by drug binding. When charged drugs are examined, the free energy factor can be split into two parts, one arising from the electrostatic contribution and the other from the non-electrostatic contribution, including hydrogen bonding and hydrophobic effects. From the temperature dependence of K_i, the enthalpy (and entropy) changes of the process are also accessible. A second important parameter is n, the number of nucleotides involved in complex formation per drug molecule. Finally ω, the cooperativity parameter , is used to define the relative tendency of a ligand to bind to an "isolated" site, in comparison to a site singly or doubly contiguous to an already occupied site.

The experimental methods for evaluating the above parameters must allow determinations of free and bound

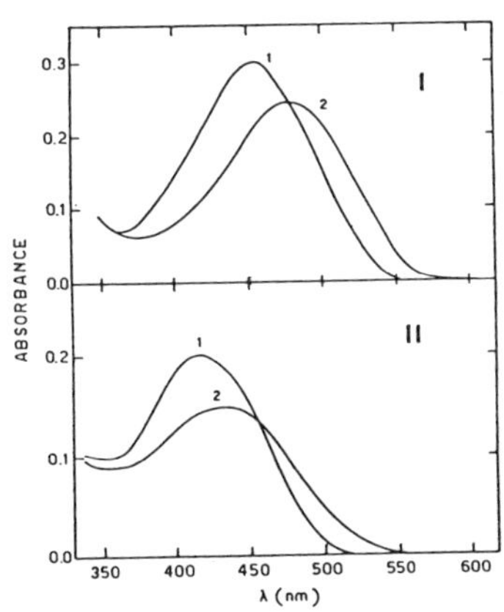

Figure 2.Spectrophotometric properties of compounds I and II free (curve 1) and bound to DNA (curve 2).

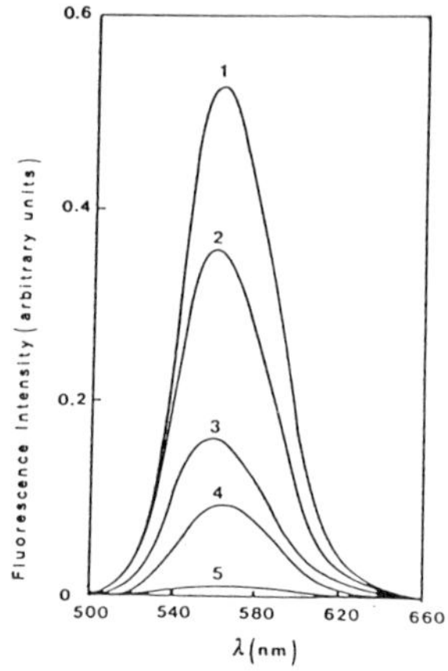

Figure 3. Fluorescence quenching of compound V by DNA. 1) free drug; 2-5) in the presence of increasing amounts of polynucleotide.

drug at equilibrium. Among various techniques most useful are equilibrium dialysis and spectroscopy. Fig. 2 and 3 illustrate characteristic changes in the ligand optical properties upon interaction with DNA. The absorption changes occurring to compounds I and II when they bind to the polynucleotide are represented in Fig. 2. They include a bathochromic shift and a remarkable hypochromism. Fig. 3 shows the fluorescence response of compound V in the presence of increasing amounts of DNA. In this case almost complete quenching of the emission intensity occurs. The results obtained by the above techniques are summarized as the Scatchard plots in Fig. 4 and 5. Using the appropriate equations (20), K_i, n and ω for the binding of the examined anthraquinones were obtained from the Scatchard plots by means of a best fit least squares program (Table I).

It is interesting to note that the intrinsic affinity values (K_i) span over a wide range (more than two orders of magnitude) as a function of anthracenedione substitution pattern. The n values are less affected, but still

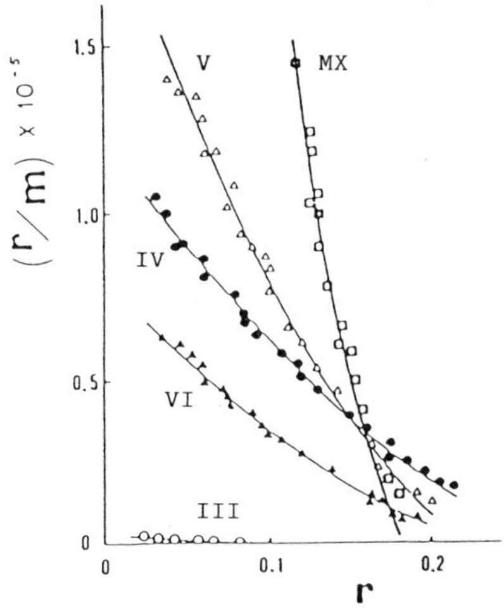

Figure 4. Non-cooperative Scatchard plots for the binding of compounds III, IV, V, VI and MX to DNA

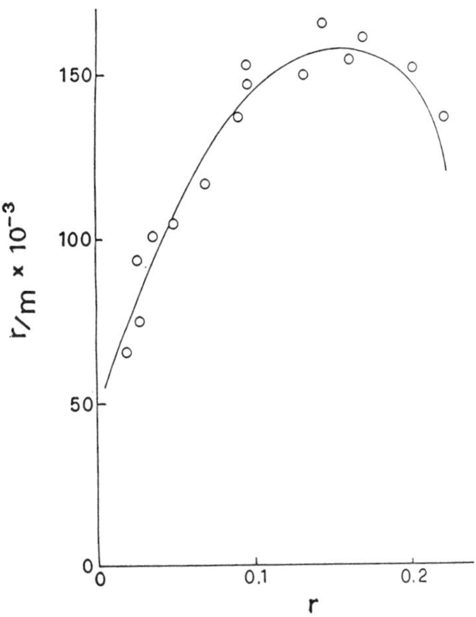

Figure 5. Cooperative Scatchard plot for the binding of compound VII to DNA

TABLE I - Thermodynamic DNA-binding parameters of anthraquinone compounds at 0.022 M ionic strength, pH 7.0, 25°C.

Compound	K_i X10^{-5}	n	ω
III	0.03	5.6	1
IV	1.2	3.1	1
V	2.0	3.8	1
VI	0.8	3.6	1
VII	0.5	4.3	32
VIII	0.2	4.8	25
IX	0.2	4.1	24
MX	10	4.0	1

the site size changes considerably from one compound to the other. Finally, 3 of them (VII-IX) possess positive cooperativity, thus preferring the formation of drug clusters along the DNA chain.

Electrostatic effects were also investigated, as compounds I-IX have positively charged side-chains, which can in principle form ionic contacts with the phosphate groups of the DNA backbone. Increasing ionic strength leads to an impairment of electrostatic interactions, with consequent decrease in DNA-affinity of the examined drugs. According to the polyelectrolyte theory applied to nucleic acids, the ionic strength dependence of K_i is related to the number of charged interactions per complex unit (m), according to the following equation (21):

$$\delta \log K_i / \delta \log |M^+| = -m\psi$$

where $|M^+|$ is the monovalent ion concentration (usually Na^+) and ψ represents the fraction of counterions thermodynamically bound to DNA (0.88 for a double-helical polynucleotide). This equation should be somewhat modified for conformational changes occurring in the macromolecule as a result of complex formation (22), but our results are not substantially affected using the above simplified formalism. Representative plots lg K_i vs. lg$|M^+|$ are shown in Fig. 6 for anthraquinones VII-IX. For the majority of compounds we found m=1, except for III and IV, for which values ranging from 0.6 to 0.7 were obtained. This fact indicates that charged side-chains at position 2 and/or 4 are able to interact efficiently with the nucleic acid, which represents a useful information for a tailored synthesis of new compounds.

By appropriate extrapolation of K_i as a function of ionic strength, the non electrostatic contribution to the binding process (K_o) can also be evaluated. Different K_o values for different compounds point to the effectiveness of the anthraquinone substitution pattern in modifying hydrophobic and hydrogen-bonding interactions with the polynucleotide matrix.

In addition to drug affinity for the macromolecule, it is very important to evaluate the binding mechanism, the structural characteristics of the DNA-ligand adducts and the conformational changes induced in the nucleic acid by drug-binding. Useful techniques in investigating the conformational aspects of anthraquinone binding to DNA in solution are viscometry, gel electrophoresis and circular dichroism.

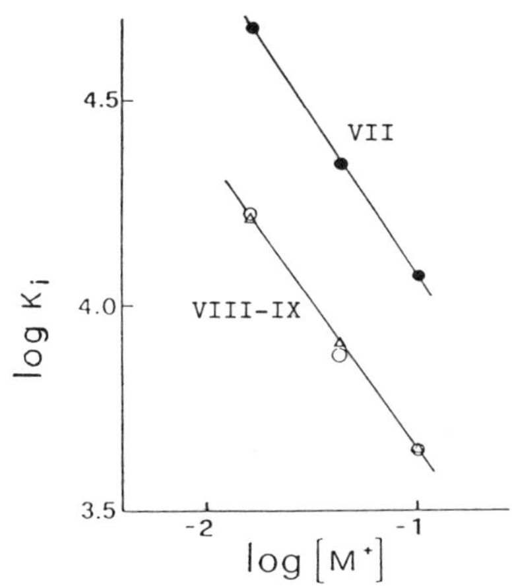

Figure 6. Ionic strength dependence of lg K_i for compounds VII-IX.

Many planar compounds interact with DNA through intercalation into the polynucleotide base pairs. This binding mechanism causes lengthening and unwinding of the double helical nucleic acid (1). To assess whether this interaction mode occurs for the series of anthraquinones examined in this paper, the electrophoretic mobility of closed supercoiled circular DNA was examined in the presence of increasing amounts of drug. Intercalation causes uncoiling of the supercoiled system, which moves slower and slower in the gel, until it is completely unwound. Further binding causes rewinding of DNA in the opposite sense, thus increasing the mobility of the complex. The electrophoretic pattern should accordingly pass through a minimum, corresponding to the unwound closed polynucleotide. When external binding occurs, the hydrodynamic properties of the macromolecule undergo slight modifications, and minor changes in mobility should be observed (23). Typical results are shown in Fig. 7 for representative anthraquinones. All were able to uncoil and further recoil pBR322 DNA except compound IV. We conclude that our anthraquinone derivatives, I-III, V-IX exhibit an intercalative binding mode, while

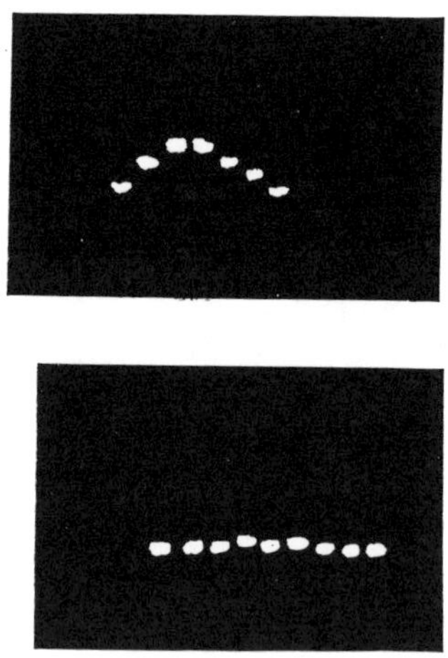

A

B

Figure 7. Electrophoretic titration of pBR 322 DNA in the presence of compound I (A) and IV (B). Drug concentration increases from left to right.

TABLE II - Unwinding ability of anthraquinone derivatives (dose range of maximal unwinding)

Compound	Dose range (ng/ml)
III	750-1100
V	50-100
VI	50-200
VII	300-500
VIII	230-300
IX	150-300

for compound IV external complexation is predominant. The DNA-unwinding ability of various derivatives is compared in Table II. It reflects the different binding affinities, site size and structural modifications for the individual compounds.

The helix-lengthening effect due to intercalation causes modifications in the specific viscosity of DNA solutions (24). The following equation can be written, which relates η_s, specific viscosity of a DNA solution when r ligand molecules are bound per DNA base, η_s^o, specific viscosity of the same DNA solution in the absence of added drug, and l, length increase per intercalated ligand:

$$(\eta_s / \eta_s^o)^{1/3} = 1+0.59\ lr$$

A plot of $(\eta_s/\eta_s^o)^{1/3}$ vs. r should give a straight line with slope 0.59 l. We found indeed linear plots for all intercalating ligands. The l values obtained for each compound are summarized in Table III.

It is worth observing that, according to the above hydrodynamic measurements, remarkably different unwinding and lengthening effects occur as a result of different substitution patterns on the same basic planar structure.

Further structural informations can be derived by circular dichroism measurements both in the DNA and in the drug absorption region. The binding process usually affects the chiroptical response of the polynucleotide in

TABLE III -DNA length increase (l) per intercalated drug

Compound:	I	II	III	V	VI	VII	VIII	IX	MX
l(nm) :	0.27	0.25	0.21	0.35	0.30	0.35	0.26	0.24	0.35

a substantial way. The observed changes in optical activity are often twofold. At low drug/DNA binding ratios intercalative compounds usually increase the rotational strength of the positive DNA band at 275 nm (25) (see for example Fig. 8 A-C). This effect is related to a shielding of base pairs due to insertion of the drug into the double helix. Interestingly, the non intercalating anthraquinone IV causes quite different CD changes, when compared to its congeners. Thus chiroptical techniques may represent an additional useful tool in gaining insight into the DNA-interaction mode of potential drugs.

For anthracenediones having high binding affinity for nucleic acids a drop of the positive CD signal is observed at high binding ratios, which eventually becomes negative before phase-separation of the complex occurs (Fig. 8B-I). If we recall that all examined compounds possess charged side-chain groups, which interact efficiently with DNA to form ion pairs, it is conceivable that extensive charge neutralization of the polyelectrolyte matrix lowers its stiffness, giving more compact forms of DNA and leading finally to collapse and aggregation of the system (26). These facts are in line with the reported transitions of DNA to tight conformations due to interactions with polycations or to excluded volume effects. Negative CD responses for the condensed forms of the nucleic acid were reported in the latter cases (27,28). The extent of condensation is found to be a function of the nature and characteristics of the examined anthraquinones. It depends upon affinity for DNA and efficiency of electrostatic bond formation. It is relevant to note that phenomena of this type have been recently proposed to represent a fundamental step in antitumor activity of mitoxantrone and ametantrone (29).

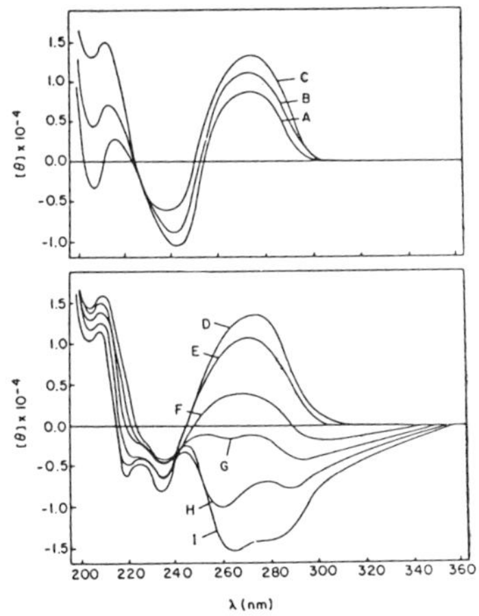

Figure 8. CD spectra of DNA in the presence of increasing amounts (from A to I) of compound I

Further insight into the stereochemistry of intercalated complexes can be gained by examining the sign of induced optical activity in the anthraquinone absorption region. Theoretical studies suggest that negative rotational strength corresponds to complex geometries in which the transition moment of the drug lies more or less parallel to the base-pair longest dimension, whereas positive CD should appear when it is oriented perpendicularly (30). Indeed the anthracycline derivatives Adriamycin and Daunomycin, for which a perpendicular geometry has been found in the solid state, show positive differential absorption in the visible. All examined intercalating anthraquinones exhibit negative induced dichroism between 450 and 550 nm (Fig. 9), suggesting that, unlike their parent compounds, they prefer a parallel geometry of binding. At high salt concentration positive induced CD responses can be observed. This might indicate that, in the absence of electrostatic interactions with DNA, the preferred orientation of the anthraquinone moiety in the intercalated complex is perpendicular to the base-pair longest dimension, as it is for antracyclines. Twisting probably occurs under low ionic strength conditions to allow electrostatic contributions to the total free energy, without substantial destabilization from repulsive steric contacts between the polynucleotide backbone and the drug substituents.

In conclusion we have shown for a number of anthraquinone derivatives that intercalative and non-intercalative binding to DNA occurs depending upon drug substitution pattern. The affinity for the nucleic acid can be appropriately modulated by changing the number and position of hydroxyl- and charged groups. Electrostatic effects contribute to the binding free energy, indicating efficient interaction between anthraquinone side chains and nucleic acid phosphate groups. The position of the substituent is shown to play a relevant role. Different structural modifications occur in DNA as a result of the binding process, depending upon ligand characteristics. These include elongation and unwinding of the polynucleotide, and compaction to condensed forms. The geometry of the intercalated complexes can be finally modified as a function of ionic strength.

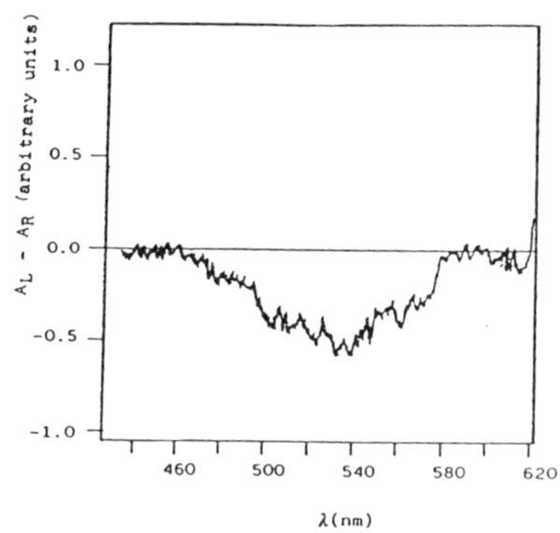

Figure 9. Induced CD in the ligand absorption region when compound VI is bound to DNA.

Biological studies, not discussed in this paper, represent the second important part of an investigation aimed at obtaining reliable structure-activity relationships. On the basis of the results from the combined physico-chemical and biological approach we are presently developing new derivatives of the anthraquinone structural type, hopefully endowed with anticancer activity, while subduing the adverse side effects of the drugs from which they are derived.

ACKNOWLEDGEMENTS

The authors are indebted to Mrs. Elvira Piaia for her expert typing of the manuscript.

REFERENCES

1. Waring, M.J. (1981) Ann. Rev. Biochem. 50, 159-193.
2. Gale, E.F., Cundliffe, E., Reynolds, P.E., Richmond M.H. and Waring M.J. (1981) in "The Molecular Basis of Antibiotic Action", 2nd ed., Wiley, London.
3. Waring, M.J. (1970) J. Mol. Biol. 54, 247-259.
4. Patel, D.J. (1980) in "Nucleic Acid Geometry and Dynamics" R.H.Sarma Ed.Oxford,Pergamon Press,185-231.
5. Di Marco, A., Arcamone, F. and Zunino, F. (1975) in "Antibiotics III. Mechanism of Action of Antimicrobial and Antitumor Agents" Corcoran J.W., F.E. Hahn Eds., 101-128.
6. Pigram, W.J., Fuller, W. and Hamilton, L.D. (1972) Nature 235, 17-19.
7. Quigley, G.J., Wang, A.H.K., Ughetto, G., Van der Marel, G., van Boom, T.H., Rich, A. (1980) Proc. Natl. Acad. Sci. USA 78, 7204-7208.
8. Huffman, D.H., Benjamin, R.S. and Bachur, N.R. (1972) Clin. Pharm. Ther. 13, 895-905.
9. Double, J.C. and Brown, J.R. (1975) J. Pharm. Pharmacol. 27, 502-507.
10. Double, J.C. and Brown, J.R. (1976) J. Pharm. Pharmacol. 28, 166-169.
11. Zee-Cheng, R.K.Y. and Cheng, C.C. (1978) J. Med. Chem. 21, 291-294.

12. Plumbridge, T.W., Knight, V., Patel D.J. and Brown, J.R. (1980) J. Pharm. Pharmacol. 32, 78-80.
13. Lu,C. and Loo, T.L. (1980) Cancer Res. 40, 1427-1433.
14. Au, W.W., Butler, M.A., Matney, T.S. and Loo, T.L. (1981) Cancer Res. 41, 376-380.
15. Palumbo,M., Antonello,C., Viano,L., Gia, O., Marciani Magno, S. (1983) Chem. Biol. Interact 44, 207-218.
16. Palumbo, M., Marciani Magno, S., (1983) Int. J. Biol. Macromol. 5, 301-307.
17. Kapuscinski, J., Darzynkiewicz, Z., Traganos, F., Melamed, M.R. (1981) Biochem. Pharmacol. 30, 231-240.
18. Lown, J.W., Hanstock, C.C., Bradley, R.D. and Scraba, D.G. (1983) Mol. Pharmacol. 25, 178-184.
19. Zee-Cheng, R.K.Y. and Cheng, C.C. (1983) Drugs Future 8, 229-249.
20. Mc Ghee, J.D. and Von Hippel, P.H. (1974) J. Mol. Biol. 86, 469-489.
21. Record, M.T., Anderson, C.F. and Lohman, T.M. (1978) Q. Rev. Biophys. 11, 103-178.
22. Wilson, W.D., Loop, G. (1979) Biopolymers 18, 3025-3033.
23. Espejo, R.T. and Lebowitz, J. (1976) Anal. Biochem. 72, 95-103.
24. King, H.D., Wilson, W.D. and Gabbay, E.J. (1982) Biochemistry 21, 4982-4989.
25. Palumbo, M., Capasso, L., Palu', G and Marciani Magno, S. (1985) J. Biosci 8, 689-697.
26. Lerman, L.S. (1971) Proc. Natl. Acad. Sci. USA 68, 1886-1890.
27. Evdokimov, Y.M., Platnov, A.L., Tikhonenko, A.S. and Varshavsky, Y.M. (1972) FEBS Letters 23, 180-184.
28. Fasman, G.D., Schaffhausen, B., Goldsmith, L. and Alder, A. (1970) Biochemistry 9, 2814-2820.
29. Kapuscinski, J. and Darzynkiewicz, Z. (1986) Proc. Natl. Acad Sci. USA 83, 6302-6306.
30. Schipper, P.E., Norden, B. and Tjerneld, F. (1980) Chem. Phys. Lett. 70, 17-21.

INTERACTION OF cis-$Ru(DMSO)_4Cl_2$ WITH DNA IN AQUEOUS SOLUTION

S. Cauci^, F. Quadrifoglio^, E. Alessio° and G. Mestroni°
^Dept. of Biochem. Biophys. and Macromol. Chem.
°Dept. of Chemical Sciences
Univ. of Trieste, Piazzale Europa 1, Trieste
Italy

INTRODUCTION

cis-$Ru(DMSO)_4Cl_2$ (cis-RDT) has been shown to possess good antitumor activity and mutagenic properties (1-5). When compared with cis-$Pt(NH_3)_2Cl_2$ (cis-PDD), a well-known and clinically used drug, the ruthenium complex shows at least a comparable activity (on LD_{50} basis). Its toxicity, however, appears remarkably lower (3), which makes this drug somewhat promising as an anticaner agent.

It has been reported that cis-PDD reacts with DNA in aqueous solution and forms covalent bonds with some of its bases, especially with guanine N7 (6). It is believed, although as yet to be proven, that these reactions are the basis for the biological action of the drug. On the other hand, complexes of Ru(II) and Ru(III) with nucleosides and nucleotides have been prepared and partially characterized (7,8). For these reasons we have undertaken a study of the interaction of cis-RDT with natural and synthetic DNAs and some of the preliminary results are presented here.

EXPERIMENTAL

cis-RDT has been synthesized as already reported (9) and recrystallized from DMSO-acetone mixtures.

Stock solutions of calf thymus DNA (Sigma, type I) were prepared dissolving the polymer in 3 mM NaCl, 1 mM phosphate buffer at pH 7.4 and diluting to a final concentration of about 8 mM_P.

The ruthenium complexes of nucleotides and polynucleotides were prepared by incubation with freshly prepared solutions of cis-RDT at different stoichiometric ratios, r_s . The reaction mixtures were kept in the dark in a water thermostatted bath with occasional manual stirring. The reaction was stopped by raising the chloride concentration to 0.5 M with 4.6 M NaCl and samples were frozen for storage.

The separation of the unreacted cis-RDT was made by ultrafiltration (stirred ultrafiltration cell Amicon, model 8010) through Diaflo Ultrafilter PM10 membrane using argon as pressuring gas. The ultrafiltration was carried out first with about 10 mL of a 1 M $NaClO_4$ solution and then with about 100 mL (10 mL x 10) of 2×10^{-4} M $NaClO_4$. The solutions recovered from ultrafiltration were analyzed by UV absorbance to determine the amount of unbound cis-RDT.

Absorbance spectra were recorded using a Cary 219 Spectrophotometer equipped with a thermostatted cuvette holder and a Haake F3 thermostat provided with a Haake PG 10 temperature programmer.

Melting curves were recorded at 257.5 nm by increasing the temperature at a rate of 0.5°C/min.

CD spectra were obtained using a Jasco J500A dichrograph connected to a Jasco DP-500N data processor and equipped with a thermostatted cell holder.

Poly(dCdG), poly(dAdT) and nucleotides were obtained from Pharmacia. pBR322 plasmid DNA (from E. coli) was obtained from Boehringer.

Restriction enzyme digestion was carried out with BamH1, EcoR1 and HindIII (Pharmacia) in 10 mM Tris-HCl, 10 mM $MgCl_2$, 100 mM NaCl at pH 8 (7.5 in the case of EcoR1). Digestions were quenched by the addition of a 1/4 digest volume of 0.25 M EDTA followed by chilling at −20°C.

Agarose gel electrophoresis was carried out in submarine horizontal slab gel (BioRad DNA SubCell). 1% gels (Agarose Serva) were prepared and run in 90 mM Tris base, 90 mM borate, pH 8.3. Gels were run at 40 V (2.6 V/cm at room temperature for about 20 hours.

The DNA was visualized by staining the gels for 1 hour in 0.5 mg/ml ethidium bromide and illuminating with UV light.

RESULTS

The neutral cis-RDT complex has an octahedral structure (10). Three out of the four DMSO ligands are bonded to ruthenium through the sulfur atom in a facial configuration while the last one is bonded through the oxygen atom (species I of Figure 1). The ruthenium atom has a formal oxidation state of +2 and a d^6-low spin electronic configuration.

The complex, in the form of bright yellow crystals, is very soluble in water. The solutions, kept in the dark, are quite stable to air oxidation. According to the labile nature of the O-bonded DMSO ligand, already evidenced by NMR studies (9,11), its dissociation in aqueous solution is very fast, so that species

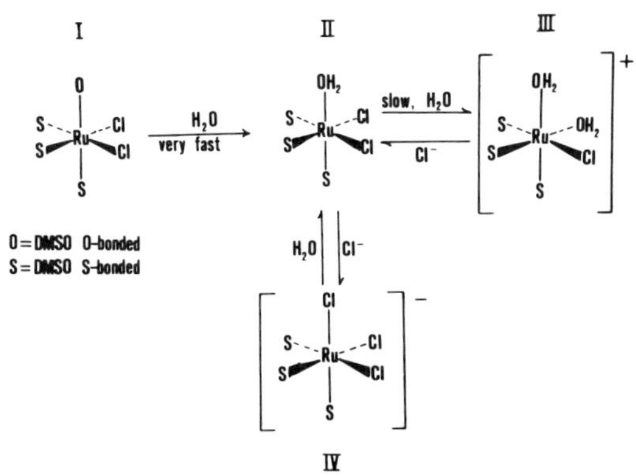

Figure 1. Scheme

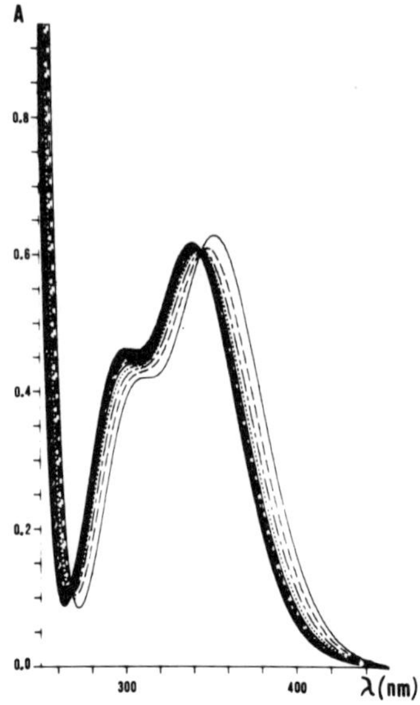

Figure 2. Absorption spectra of cis-RDT in water at 25°C as a function of time. The spectra were recorded with an interval time of about 30 min. The right spectrum is at time zero.

of Figure 1 is the one present as soon as the complex is dissolved in water. Complex II, as clearly evidenced by conductivity measurements (9), undergoes a slow dissociation of a chloride ion to give the cationic species III. The slow chloride dissociation is also confirmed by the change of the electronic spectrum with time shown in Figure 2. The presence of two isobestic points is strongly in favor of the presence in solution of only two species in equilibrium in the time range explored. Conductometric and spectral measurements carried out in water showed that the equilibrium dissociation of the first chloride ion is attained within about 12 hours at 25°C and within about 3 hours at 37°C. Any further, significant, dissociation is not evidenced, even after seven days at 25°C, in contrast to what is stated elsewhere (9).

The equilibrium between species II and III is affected by the Cl^- concentration. We have measured such equilibrium at two significant Cl^- concentrations, namely 3 mM and 150 mM, which roughly correspond to intra- and extracellular conditions, respectively. At the lowest Cl^- both kinetic and thermodynamic behaviors are very similar to those found in pure water, whereas in 150 mM Cl^- the reaction II ⟶ III is almost completely inhibited. This result might have some relevant biological significance. Increasing the Cl^- concentration up to 1.5 M a negative complex is formed (IV) (12).

Figure 3 shows a typical absorption spectrum of calf thymus DNA after incubation with cis-RDT. The excess of absorbance shown beyond 300 nm with respect to untreated DNA is due to the formation of some kind of Ru-DNA complex. The treatment of the DNA solutions after incubation, described in the experimental section, rules out any possible presence in the solution of positively charged complexes bound only electrostatically to DNA chains.

The changes in the spectra produced by the reaction can be used to monitor the time course of the complexation. The A_{230}/A_{258} ratio as a function of the incubation time, reported in Figure 4, shows that a first saturation level of bound ruthenium is attained within 2-3 days, but that 90% of such saturation is already reached within 10 hours. For calf thymus DNA the value of ruthenium bound ($r_b=Ru_b/DNA$, determined as described in the experimental section) corresponding to 3 days of incubation at $r_s=1$, 25°C and 3 mM NaCl, 1 mM phosphate pH 7.4, amounts to about 0.25.

A chemical modification of the complex formed with DNA seems to be responsible for the spectral changes registered at longer incubation times as the value of r_b is not appreciably increased.

Figure 5 shows the CD spectra of both native and heat denatured calf thymus DNA before and after incubation. The data show that the B conformation of the polynucleotide chain is basically unaltered by the ruthenium bonding, the only observable differences in the native CD spectra being the slight modification of the amplitudes of both positive and negative bands

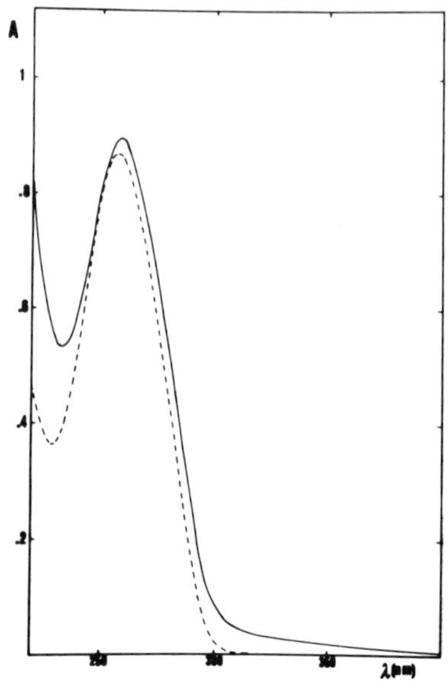

Figure 3. Absorption spectra of calf thymus DNA before (----) and after 7 days incubation at 25°C with cis-RDT (——).

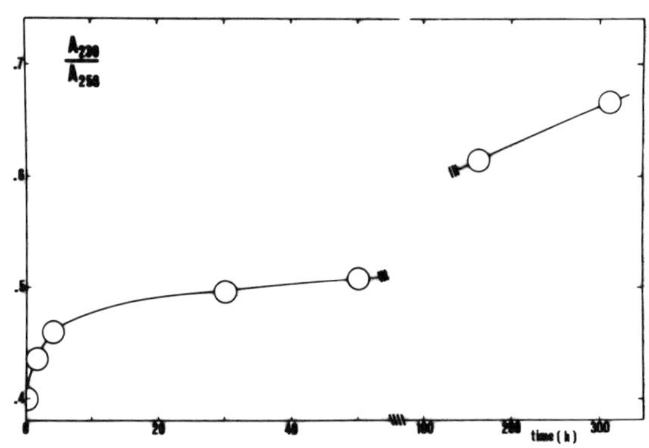

Figure 4. A_{230} /A_{258} as a function of incubation time for Ru-DNA complexes.

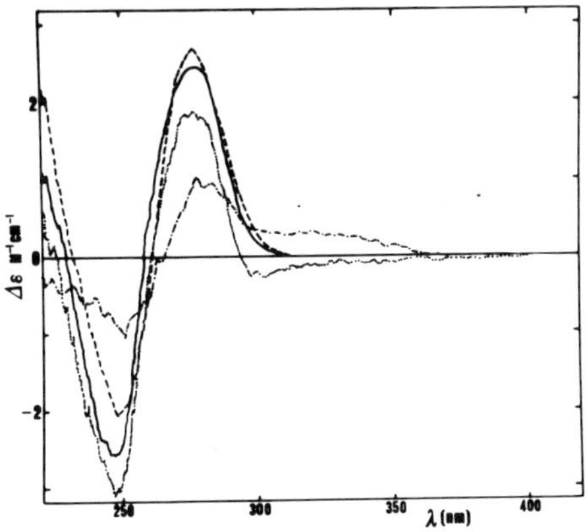

Figure 5. CD spectra of native (....) and heat denatured (-.-.) calf thymus DNA complexed with cis-RDT. For comparison the spectra of untreated native (——) and heat denatured (----) DNA are reported. All measurements performed at 25°C.

and the appearance of a low intensity negative band at around 300 nm. The differences are more marked in the heat denatured state.

The lack of disruption of the B conformation in calf thymus is also confirmed by the melting experiments carried out with DNA-Ru complexes at different degrees of bound ruthenium, r_b . Surprisingly enough the B structure of DNA is stabilized by the cis-RDT binding and the increase of T_m with the incubation time parallels quite well that of the degree of saturation (Figure 4 and 6). The melting curves resulted not reversible upon cooling. This behavior is different from that found in the case of cis-PDD whose reaction with DNA destabilizes the B helix (13).

Figure 7 shows the Scatchard plots corresponding to the binding of ethidium bromide to calf thymus DNA before and after incubation with cis-RDT. The measurements have been performed in ionic strength conditions such that only the intercalative binding of ethidium is possible. The results show a decrease in the number

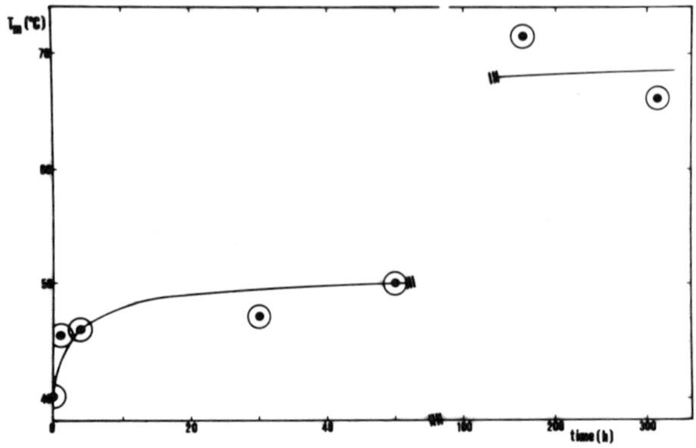

Figure 6. Melting temperatures of DNA-Ru complexes in 2 x 10^{-4}M $NaClO_4$, as a function of incubation time.

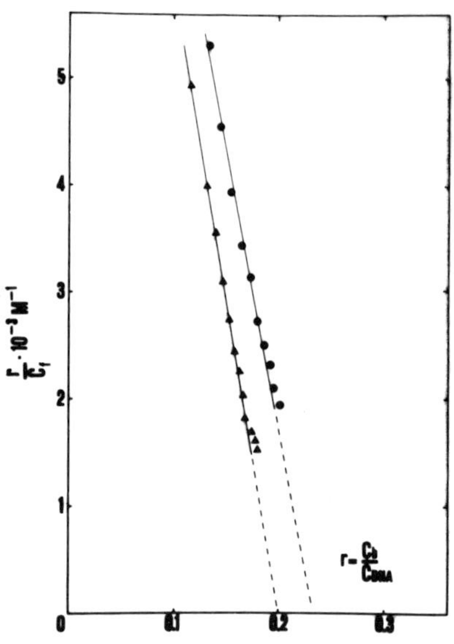

Figure 7. Scatchard plots for ethidium bromide binding to DNA untreated (●) and treated (▲) with cis-RDT. Solvent: 0.5 M NaCl, T=25°. The amount of ethidium bromide was determined by absorption measurements.

of binding sites following ruthenium bonding, but no effects on the apparent binding constant are found.

Although several complexes with purines and pyrimidines have been prepared in the past (7,8) we have tried to incubate directly 5'GMP and 5'AMP with the ruthenium complex in the same experimental conditions used for DNA solutions. After 10 days of incubation at 25°C and r_s =1, the final mixture was eluted with water through a Biogel P2 column. Several peaks were collected in each case, some of which presented dichroic bands at wavelengths higher than those corresponding to the electronic transition of the base itself and compatible only with ruthenium-base complexes (data not shown). Preliminary results show that 5'CMP also forms stable complexes with cis-RDT.

The multiplicity of dichroic spectra obtained after the reaction of a nucleotide with cis-RDT suggests an analogous multiplicity in the nature of the complexes formed, which can differ either in the number of purines or pyrimidines attached to the ruthenium and/or in the atom used by the bases for the reaction with the metal. All these complexes in aqueous solution show a fairly good stability to air oxidation which appears, however, lower than that exhibited by the Ru-DNA complexes.

The variety of compounds which nucleotides and cis-RDT can form makes it difficult to identify the complexes formed with DNA. This can be done, however, by enzymatic hydrolysis of the Ru-DNA chain followed by separation and identification of the base-ruthenium complexes. This work is under way in our laboratory, but some useful information in this respect has already been obtained by studying ruthenium complexes of synthetic DNAs with definite base sequence.

Figure 8 shows the change with time of the CD spectrum of poly(dCdG) dissolved in $2x10^{-4}$ M $NaClO_4$ and in the presence of an equimolar amount of cis-RDT. The data show that the polymer undergoes a B to Z transition with time and that the conversion is almost complete within 8-10 hours at 37°C. As the transformation is not reversed by the addition of EDTA, it is safe to conclude that the transition is not due to the presence of trace amounts of di- or trivalent cations (14),

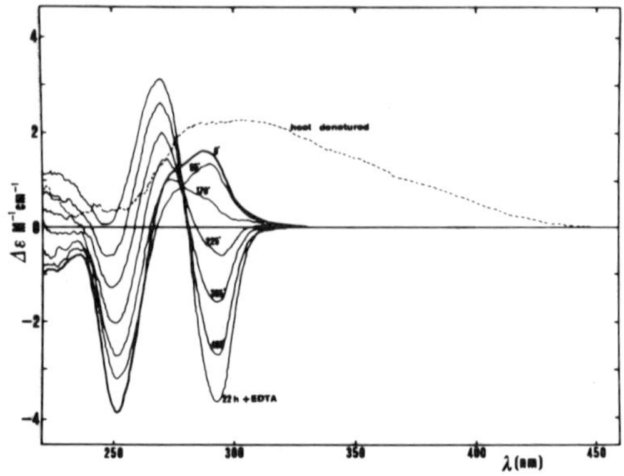

Figure 8. CD spectra, registered at different times, of poly(dCdG) ($c=1.18 \times 10^{-4}$ M) dissolved in 2×10^{-4} M $NaClO_4$ and in the presence of equimolar amount of cis-RDT. At t=22 h EDTA was added to a final concentration of 2×10^{-4} M and then the solution was heated up to 90°C to produce the denatured form whose CD spectrum was recorded at room temperature.

but to the binding of the ruthenium complex. This result strongly suggests that N7 of guanine residues is the reaction target for cis-RDT. In fact, it has already been shown that substitutions at N7 of guanine rings by bulky substituents stabilize the syn conformation of the purine, which in turn stabilizes the Z form of the alternating polymer (15-19).

Heating the complexed poly(dCdG) to 90°C in the presence of the unreacted cis-RDT produces a denatured state whose CD spectrum is also shown in Figure 8, and which does not renature upon cooling.

An analogous type of experiment carried out on poly(dAdT) (dissolved in 0.1 M $NaClO_4$) produced no change in the B type CD spectrum with time (at least within 26 hours), although binding of cis-RDT was detected by the ultrafiltration method. Heating up the solution to 90°C produces a denatured form whose CD spectrum shows bands beyond 300 nm (data not shown). Also in this case the renaturation of the sample does not occur upon cooling. However, when the denaturation-renaturation cycle is done by monitoring the absorbance and by limiting the rise in temperature to the minimum value required for the complete denaturation, it is possible to obtain, upon cooling, full renaturation. Successive cycles performed in the same way gave progressive reductions of the fraction of renatured polymer. After 4-5 cycles, renaturation of poly(dAdT) was completely abolished.

It has been reported that progressive binding of cis-PDD to circular, supercoiled DNA shows an effect similar to that produced by an intercalator, namely relaxation and then reversion of supercoiling (20, 21). Such an effect has not been obtained with the system pBR322 plasmid DNA/cis-RDT. In fact, no change in the rate of migration of the supercoiled form through a 1% agarose gel, with the same incubation time, was noticed.

cis-RDT and cis-PDD also differ in the capability of protecting restriction sites in the DNA chains. In fact, EcoR1, BamH1 and HindIII did not show any appreciable reduction of their activity toward their relative, unique sites in the pBR322 chain previously incubated with cis-RDT. On the other hand cis-PDD has been reported to protect restriction sites containing G stretches (21,22). Following treatment with cis-RDT, however, a marked relative increase of the nicked form of the plasmid was noticed.

DISCUSSION

The experimental results reported in this communication allow us to draw the following conclusions:

1. cis-RDT forms covalent bonds with DNA. This is demonstrated by the retention of the long wavelength absorption band after ultrafiltration of the Ru-DNA complex with 1 M $NaClO_4$ which should remove any electrostatic interaction between negatively charged DNA and the positively charged ruthenium complex.
2. For calf thymus DNA and at stoichiometric ratios Ru/DNA = 1, the degree of complex bound (r_b) reaches a value of about 0.2 within 10-12 hours of incubation at 25°C.
3. Within this time interval the reactive forms of the ruthenium complex should be species II and III of Figure 1. In both cases the final product should

bear a positive charge as dissociation of Cl^- is not prevented by the reaction with DNA (on the contrary, it should be favored by the negative charge of the chain).

4. A very likely target for the reaction on the DNA chain is the N7 atom of guanine. This position has already been demonstrated as the binding site for other ruthenium complexes (23-24) but in our case the suggestion is strongly supported by the experiments carried out with poly(dCdG). In fact, bulky substituents in N7 or C8 of the purine ring are known to drive the alternating polymer toward the Z conformation (15-19) and C8 does not appear as a possible binding site for the ruthenium complex.

5. N7 of adenine rings is also a possible binding site for ruthenium. In fact poly(dAdT) reacts with cis-RDT and a consequence of the reaction is the stabilization of its B structure. Had the adenine reacted with N1 (or N6), which is the preferred binding site (25), the consequence would have been the destabilization of the ordered structure.

6. On the same grounds, cytosine and thymine should be discarded as possible reaction targets.

7. The increase in T_m and the very small changes in the CD spectrum of calf thymus DNA and of poly(dAdT) together with the absence of influence on the mobility of the supercoiled plasmid DNA following reaction with cis-RDT suggest that the bulky metal complex fits well into the major groove of the B conformation after coordination with N7 of the purine rings. The positive charge of the complex and the possibility to form hydrogen bonds with neighboring groups can explain the overall increased stability of the structure.

8. The marked increase in the amount of the nicked form of the plasmid DNA after reaction of pBR322 with cis-RDT is indicative of a weakening of the phosphate ester bond due to complex binding.

9. Long incubation times or heating at high temperatures produce additional effects, whose nature is not yet clear.

In all these aspects, however, work is in progress in our laboratory.

NOTE. After the Siena Meeting the authors became aware of the paper published by M.J. Clarke, B. Jansen,

K.A. Marx and R. Kruger ((1986) Inorg. Chim. Acta 124, 13-28) on the binding of $[Ru(NH_3)_5H_2O]^{+2}$ to DNA. Most of our results and conclusions are similar to theirs. The only exception is the destabilization of the DNA by the complex they used.

REFERENCES

1. Giraldi, T., Sava, G., Bertoli, G., Mestroni, G. and Zassinovich, G. (1977) Cancer Res. 37, 2662-2666.
2. Sava, G., Giraldi, T., Mestroni, G. and Zassinovich, G. (1983) Chem. Biol. Interact. 45, 1-6.
3. Sava, G., Zorzet, S., Giraldi, T., Mestroni, G. and Zassinovich, G. (1984) Eur. J. Cancer Clin. Oncol. 20, 841-847.
4. Monti-Bragadin, C., Ramani, L., Samer, L., Mestroni, G. and Zassinovich, G. (1975) Antimicrob. Agents Chem. 7, 825-827.
5. Monti-Bragadin, C., Tamaro, M. and Banfi, E. (1975) Chem. Biol. Inter. 11, 469-472.
6. Lippard, S.J. (1982) Science 218, 1075-1082.
7. Farrell, N. and de Oliveira, N.G. (1982) Inorg. Chim. Acta 66, 161-163.
8. Grigoratos, A. and Katsaros, N. (1985) Inorg. Chim. Acta 108, 41-45.
9. Evans, I.P., Spencer, A. and Wilkinson, G. (1973) J. Chem. Soc. Dalton Trans. 204-209.
10. Mercer, A. and Trotter, J. (1975) J. Chem. Soc. Dalton Trans. 2480-2483.
11. Farrell, N. and de Oliveira, N.G. (1980) Inorg. Chim. Acta 44, 1255-1256.
12. McMillan, R.S., Mercer, A., James, B.R. and Trotter, J. (1975) J. Chem. Soc. Dalton Trans. 1006-1010.
13. Macquet, J.P. and Butour, J.L. (1978) Biochimie 60, 901-914.
14. Woisard, A., Guschlbauer, W. and Fazakerley, G.V. (1986) Nucleic Acids Res. 14, 3515-3519.
15. Mercado, C.M. and Tomasz, M. (1977) Biochemistry 16, 2040-2046.
16. Sage, E. and Leng, M. (1980) Proc. Natl. Acad. Sci. USA 77, 4597-4601.
17. Sage, E. and Leng, M. (1981) Nucleic Acids Res. 9, 1241-1250.

18. Santella, R.M., Grunberger, D., Weinstein, B. and Rich, A. (1981) Proc. Natl. Acad. Sci. USA 78, 1451-1455.
19. Malfoy, B., Hartmann, B. and Leng, M. (1981) Nucleic Acids Res. 9, 5659-5668.
20. Cohen, G.L., Bauer, W.R., Barton, J.K. and Lippard, S.J. (1979) Science 203, 1014-1016.
21. Ushay, H.M., Tullius, T.D. and Lippard, S.J. (1981) Biochemistry 20, 3744-3748.
22. Cohen, G.L., Ledner, J.A., Bauer, W.R., Ushary, H.M., Caravanat, C. and Lippard, S.J. (1980) J. Am. Chem. Soc. 102, 2487-2488.
23. Clarke, M.J. and Taube, H. (1974) J. Am. Chem. Soc. 96, 5413-5419.
24. Rubin, J.R., Sabat, M. and Sundaralingam, M. (1983) Nucleic Acids Res. 11, 6571-6586.
25. Clarke, M.J. (1978) J. Am. Chem. Soc. 100, 5068-5075.

STUDIES ON THE INTERACTION BETWEEN RIBOSOME AND ELONGATION FACTOR 2 BY FLUORESCENT LABELING OF THE DIPHTAMIDE RESIDUE

A. Giovane, L. Quagliuolo, L. Servillo, C. Balestrieri
Istituto di Chimica Biologica, I Facoltà di Medicina
Via Costantinopoli, 16 - 80138 Napoli

The elongation factor 2 (EF-2) catalyzes translocation, which is the last step of the elongation cycle in the course of protein synthesis. The factor is the only known protein that is ADP-ribosylated by diphtheria toxin in the presence of NAD. The target of this reaction is a post-translationally modified histidine residue, called diphthamide. ADP-ribosylated EF-2 is unable to translocate, but still retains the ability to form a ternary complex with GDP and ribosome (1). We demonstrated that diphtheria toxin is able to recognize as a substrate an NAD analog, nicotinamide 1,N^6 ethenoadeninedinucleotide (ε NAD), and to transfer the fluorescent ε ADP-ribose moiety onto the EF-2 diphthamide residue (2). Therefore, it was possible to label the EF-2 with this fluorescent probe by an efficient and selective enzymatic mechanism which allowed us to confidently attribute the molecular localization of the fluorophore at the level of the diphthamide residue. The labeled protein was then subjected to fluorescence quenching studies, alone and in the presence of GDP and ribosomes, in order to obtain elucidation on the topology of the diphthamide residue in the free protein and in the ternary complex.

MATERIALS AND METHODS

EF-2 was prepared from human placenta as described in Giovane, et al. (2). Also described in (2) is the purification of ε NAD, which was obtained from P.L. Biochemical, and the labeling of EF-2 by diphtheria toxin in the presence of ε NAD. Ribosomes were prepared from human placenta according to Palmiter (3). The 80 S particles, deficient in elongation factors, were obtained as described by Merrick (4).

Fluorescence measurements were taken at 20°C on a Perkin Elmer MPF-44B spectrofluorometer. The fluorescence quenching was monitored by a fluorescence intensity decrease at 415 nm, using an excitation light of 320 nm. The quenching measurements were made with 1 μM ε ADP-ribosylated EF-2, 50 μM GDP, 1 μM human placenta 80 S ribosomes in 20 mM Tris-HCl, pH 7.5, 60 mM KCl, 10 mM $MgCl_2$, 6 mM mercaptoethanol.

RESULTS AND DISCUSSION

Figure 1 shows fluorescence quenching data for free ADP, ADP-ribosylated EF-2 and ADP-ribosylated EF-2 in the presence of GDP, ribosomes or both.

The data are linearly correlated, thus indicating that the quenching, either by iodine or by acrylamide, follows the Stern-Volmer relationship (5). It is evident from the slope of the line in Figure 1 that there is only a small difference in the iodide accessibility to the fluorophore between the free ε ADP and the ε ADP-ribosylated EF-2. Namely, the Stern-Volmer constant decreases by a 1.3 factor. Moreover, the quenching data on the ε ADP-ribosylated EF-2, in the presence of GDP or ribosomes, superimpose onto those of the protein alone. These results lead to the conclusion that, in the ε ADP-ribosylated EF-2, the ε adenine is most probably on the protein surface, exposed to the quencher almost as in the free fluorophore; neither the presence of GDP nor that of the ribosomes modifies this situation. This is not an unexpected finding: diphthamide is a result of a complex post-translational modification, involving five proteins acting on a histidine residue (6). It is, therefore, hardly conceivable that, in the free protein, this residue is buried inside the protein

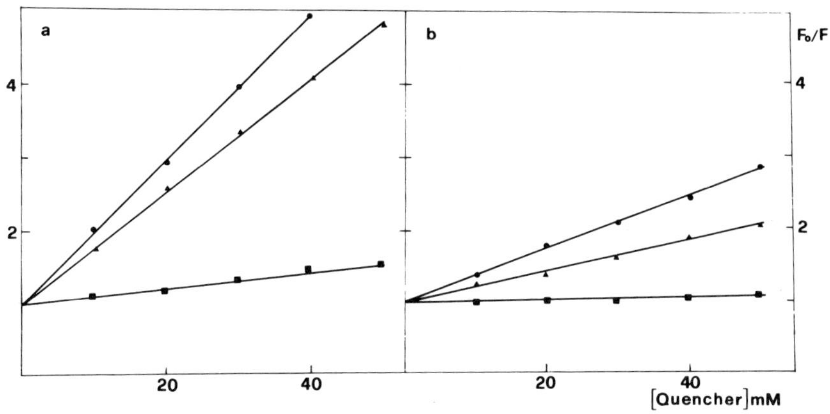

Figure 1. Stern-Volmer plots of iodide (a) and acrylamide (b) quenching of free ε ADP (-●-●-) ; ε ADP-ribosylated EF-2 (-▲-▲-) ; ε ADP-ribosylated EF-2 in the ternary complex with GDP and ribosome (-■-■-). The quenching data relative to the ε ADP-ribosylated EF-2 in the presence of GDP or of ribosomes were superimposed onto those from the protein alone.

matrix. The converse holds true when the ε ADP-ribosylated EF-2 is in the presence of both GDP and ribosomes. The slope steeply decreases (by a 7.1 factor with respect to the ε ADP-ribosylated EF-2 alone) indicating that the fluorophore transfers to an environment not easily reached by the quencher.

The results obtained with acrylamide as a quencher (Figure 1-b) are similar to those obtained with iodide. The acrylamide resulted to be less effective than iodide for the free ε ADP and for the ε ADP-ribosylated EF-2; but this result was expected because of the larger dimension of the acrylamide molecule compared to that of iodide, and because of the asymmetry of acrylamide, This latter property renders the collisions with the fluorophore less effective for the quenching process. However, in the quenching experiments with acrylamide on the ADP-ribosylated EF-2, in the presence of GDP and ribosomes, the Stern-Volmer constant decreases by a still higher factor (about 8.2), with respect to the experiment without GDP and ribosomes, than those performed with iodide as a quencher. These results leave no doubt that εADP-ribosylated EF-2, in the presence of GDP and ribosomes, is able, like EF-2, to form a ternary complex.

The quenching experiments lead us to suppose that in EF-2 the diphthamide residue, to which the εADP-ribose is transferred by diphtheria toxin, is on a part of the protein surface which is engaged in the binding with the ribosome. As a consequence of complex formation, the diphthamide is no longer exposed to the solvent and, therefore, is not easily reached by the quencher. The observation that there is no difference in quenching efficiency, both with iodide and with acrylamide, between the εADP-ribosylated EF-2 alone and with ribosomes, leads us to conclude that the presence of GDP is necessary for complex formation. On the other hand, the εADP-ribosylated EF-2 does not have a different behavior in the presence of GDP. Many authors hypothesize (7-9) that a conformational change takes place as a consequence of the guanyl nucleotides binding to the EF-2. Our results indicate that, if a conformational change occurs, it does not involve the diphthamide residue which retains the same exposure to solvent after binding.

REFERENCES

1. Bermek, E. (1978) Progress in nucleic acid research and molecular biology 21, 63-100.
2. Giovane, A., Balestrieri, C., Quagliuolo, L., Servillo, L. (1985) FEBS Lett. 191, 191-194.
3. Palmiter, R.D. (1974) Biochemistry 13, 3606-3615.
4. Merrick, W.C. (1979) Methods Enzymol. 60, 108-123.
5. Stern, O. and Volmer, M. (1919) Phys. Z. 20, 183-190.
6. Van Ness, B.G., Barrowclough, B. and Bodley, J.W. (1980) FEBS Lett. 120, 4-6.
7. Sperti, S., Montanaro, L. and Mattioli, A. (1971) Chem. Biol. Interact. 3, 141-148.
8. Nygard, O. and Nilsson, L. (1984) Eur. J. Biochem. 140, 93-96.
9. Nilsson, L. and Nygard, O. (1985) Eur. J. Bioche. 148, 299-304.

SECTION III

IMMUNOLOGICAL RECOGNITION AND DEVELOPMENT OF SYNTHETIC VACCINES

FINE SPECIFICITY OF MONOCLONAL ANTI-LYSOZYME ANTIBODIES AND THE THREE-DIMENSIONAL STRUCTURE OF A LYSOZYME-ANTIBODY COMPLEX

A.G. Amit, M. Harper, F. Lema, R.A. Mariuzza, S.E.V. Phillips*, R.J. Poljak

Département d'Immunologie, Institut Pasteur, 75724 Paris Cedex 15, France, *Astbury Department of Biophysics, University of Leeds, Leeds LS2 9KT England

Antigen-antibody interactions such as they occur in the course of adaptive immune responses in vertebrates are highly specific. Since proteins are amongst the most frequent and diverse antigens encountered by immune systems, a large number of immunochemical studies have been made using model proteins to characterize the precise interactions between antigens and antibodies in the course of immune responses (reviewed in Benjamin et al., 1). With the aim of attaining atomic resolution in the description of those interactions, we have chosen hen egg-white lysozyme (HEL) as an antigen. HEL is a small, thoroughly characterized protein, easy to obtain and purify and very immunogenic in the BALB/c strain of mice used in this work. HEL variants, with amino acid sequence differences scattered all along their sequences occur in the eggs of different gallinaceous birds. The three-dimensional structure of HEL has been determined (2) as well as those of turkey (3), human (4) and phage T4 lysozyme (5).

Antigenic determinants recognized by polyclonal and

monoclonal anti-HEL antibodies raised principally in rabbits and in the A/J and BALB/c strains of mice have been analyzed in several laboratories by immunochemical techniques (reviewed in Benjamin et al., 1). However, these studies are insufficient to map the entire antigenic determinant recognized by an antibody. Other equally important questions remain to be answered. For example, are the antigenic determinants made by continuous features of the amino acid sequence of the antigen or are they discontinuous? Are there conformational changes in the antibody that could transmit a signal to initiate effector functions? Are there conformational changes in the antigen after reaction with the antibody? How can variations in the amino acid sequence of an antigen affect its recognition by specific antibodies? Providing answers to these questions requires the study of the three-dimensional structure of an antigen-antibody complex, a study that can be carried out by X-ray diffraction techniques using monoclonal antibodies of predefined specificity, obtained by cellular hybridization techniques (6).

EXPERIMENTAL

The preparation and characterization of anti-HEL monoclonal antibodies has been described (7) as well as the preparation and crystallization of the complex between Fab D1.3 and HEL (8). The crystals are monoclinic, space group $P2_1$, with $\underline{a}$ = 55.6 Å, $\underline{b}$ = 143.4 Å, $\underline{c}$ = 49.1 Å, β = 120.5°, with one complex in the asymmetric unit. They grow from 15-20% polyethylene glycol 8000 solutions at pH 6.0 reaching dimensions of up to 1.2 x 0.5 x 0.3 mm.

X-ray intensity measurements were performed using a single counter, four-circle automatic diffractometer. Possible heavy atom derivatives were screened by taking low angle (5°-9°) precession photographs of substituted crystals to detect intensity changes, and by subsequent intensity measurements in the diffractometer. CuKα radiation was used in all these esperiments. After refinement of heavy atom parameters and native protein phases, Fourier maps of the electron density of the crystals were calculated and plotted on transparent sheets for a preliminary study. A graphics display system (PS 300, Evans & Sutherland) and the program Frodo (9) were used to interpret maps at 6 Å and at 2.8 Å resolution. The amino acid sequences of the variable regions of the heavy chain (V_H) and of the light chain (V_L) determined from the cDNA sequences (10) were used to interpret the 2.8 Å resolution map and to build a threedimensional model. The model was submitted to crystallographic refinement; the error of its atomic positions can be estimated (11) to be approximately ± 0.6 Å.

RESULTS

The monoclonal anti-HEL antibodies obtained by cell hybridization techniques are listed in Fig. 1. The characterization of their reactivity and fine specificity using a panel of 9 lysozymes allows to subdivide them into 10 groups. Although the antibodies included within a given group share overall fine specificity, they differ from each other in physicochemical properties, indicating that they do not derive from repeat isolates of the same clone. All the monoclonal anti-HEL antibodies listed in Figure 1 are IgG1,kappa. Pairs of antibodies that recognize different, non-overlapping areas of the antigen's surface, such as D1.3 and D74.3 or D1.3 and D44.2, give rise to ternary Fab-HEL-Fab complexes, indicating that the reaction with one antibody combining site does not alter the conformation of the antigen in a way such that it cannot be recognized by a second, non-overlapping antibody combining site.

The association constants of some of the antibodies have been determined using HEL as the ligand. The following values were found: D1.2, 5.3 (± 1.0) x 10^7 M^{-1}; D1.3(Fab), 4.5 (± 0.6) x 10^7 M^{-1}; D11.10, 1.7 (±0.5) x 10^7 M^{-1}; D44.2, 1.8 (±0.5) x 10^7 M^{-1}; D74.3, 1.5 (± 0.4)x 10^7 M^{-1}; D122.13, 7.9 (± 1.5) x 10^6 M^{-1}. Heteroclitic antibo-

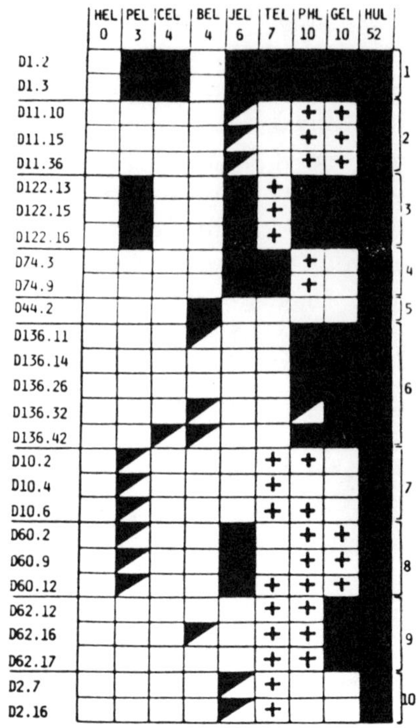

Figure 1. Direct binding of monoclonal antibodies to a panel of avian and human lysozymes and humanolysozyme (adapted from Harper et al., 1986). Binding was determined by enzyme-linked immunosorbant assays (ELISA). Egg-white avian lysozymes were obtained from hen (HEL), partridge (PEL), California quail ((CEL), bobwhite quail (BEL), Japanese quail (JEL), turkey (TEL), pheasant (PHL) and guinea hen (GEL). HUL is human milk lysozyme. The number under each lysozyme label gives that of the amino acid sequence differences with HEL. The 27 monoclonal antibodies assayed are listed in the leftmost column (D1.2 to D2.16). The relative level of binding of a monoclonal antibody to cross reactive lysozymes is shown as a percentage of the binding to HEL: (+): 125% (heteroclitic binding); blank: 68-125% (equivalent binding); black and white: 26-67% (intermediate binding); black: 0-25% (negative binding).

dies such as those in the D11 group (see Figure 1) bind heterologous antigens with slightly higher association constants.

Analysis of the amino acid sequences of the different avian lysozymes of the test panel shown in Figure 1, indicates that a single amino acid replacement, that of Gln 121 by His, is enough to prevent recognition of the antigen by antibodies D1.2 and D1.3. Consequently, these antibodies recognize an antigenic determinant that includes position 121.

The three-dimensional model of the FabD1.3-HEL complex shows that the interacting antigen-antibody surfaces are very complementary and give a dense packing of amino acid side chains and even main chain to main chain contacts. These contacts extend over a large area of about 20 x 30 Å. The lysozyme residues which are in contact ($\leq$ 4 Å) with the antibody combining site are: Asp18, Asn19, Arg21, Gly22, Tyr23, Ser24, Leu25, Asn27, Lys116, Gly117, Thr118, Asp119, Val120, Gln121, Ile124 and Leu129. These sixteen amino acid residues of lysozyme contact seventeen residues of the antibody combining site: V_L residues His30, Tyr32, Tyr49, Tyr50, Phe91, Trp92, Ser93, and V_H residues Thr30, Gly31, Tyr32, Trp52, Gly53, Asp54, Arg99, Asp100, Tyr101 and Arg102.

A quantitative comparison of the Fab-bound HEL with free HEL shows that the atomic coordinates of their alpha-carbons (C_α) differ by about 0.64 Å. A similar comparison of the complexed Fab1.3 with those of other Fabs (12-14) indicates no major change in the tertiary structure of FabD1.3. However, the Fab appears in the most extended conformation of any Fab analyzed to this date, with the angle between the V_H-V_L axis and the C_H1-C_L axis close to 180°.

DISCUSSION

It may appear perhaps surprising that no major conformational change (> 2.0 Å) can be observed in the antigen. Very small movements (~ 1 Å) of some of the antigen side chains cannot be ruled out at the current 2.8 Å resolution of this structure determination. Similarly, no major conformational change appears to take place at the antibody combining site.

The tightly packed antigen-antibody interface excludes solvent at most places. Many of the antibody side chains (9 out of 17) making contact with the antigen are aromatic. All the CDRs of the antibody make contacts with the antigen. However, two of the contacting residues, V_H Thr30 and V_L Tyr49 correspond to the less variable "framework" regions (15). The H chain makes more contacts than the L chain, and in particular its CDRs makes more contacts than any of the other CDRs. Of the 10 hydrogen bonds made between antigen and antibody, 5 are made by V_H CDR3 whose contacting residues, Arg99, Asp100, Tyr101 and Arg102 are enclosed by a D segment sequence. This shows that the D-segment diversity (reviewed by Tonegawa, 16 and Honjo, 17) is functionally significant. Positions at which variable sequences arising from imprecise joining of V_L and J_L, V_H and D, and D and J_H segments could occur do not make contacts with the antigen in antibody D.1.3.

The fact that antibody D1.3 does not bind ($K_A < 1x10^5$ M^{-1}) lysozymes that have an amino acid replacement at position 121 can be explained by the fact that the residue Gln121 in HEL makes many close contacts with the antibody, including a hydrogen bond by its side chain (N_ε)H group. Although replacement of Gln121 by His, as in the egg-white lysozymes from turkey, partridge, and California quail, would not seem to give rise to a problem of steric hindrance, factors such as 1) a change of conformation in the main chain, 2) a different orientation of His stabilized by local interactions or 3) a charge in the imidazole side chain, could affect complex formation with antibody D1.3. The solvent excluded area (18) of the antigen (748 Å^2) and of the antibody (690 Å^2) after complex formation is of an order of magnitude greater than observed in other protein-protein complexes (19) that have been analyzed. It thus appears that specificity in antigen-antibody reactions is achieved by close interactions over a large surface area. These interactions involve little or no conformational change, in general agreement with the "lock and key" proposal formulated by Ehrlich in 1900.

Acknowledgements. We thank M. Verhoeyen for making the amino acid sequences of V_H and V_L D1.3 available to us before publication, C. Wilmot for solvent-accessible surface calculations, F. Gauthier for expert secretarial help and the Laboratoire d'Utilisation du Rayonnement Synchrotron, C.N.R.S. for the use of a graphics system. This research was supported by grants from the Institut Pasteur C.N.R.S., INSERM and a short term EMBO fellowship (ASTF 4777) to S.E.V. Phillips.

REFERENCES

1. Benjamin, D.C., Berzofsky, J.A., East, I.J., Gurd, F. R.N., Hannum, C., Leach, S.J., Margoliash, E., Michael, J.G., Miller, A., Prager, E.M., Reichlin, M., Sercarz, E.E., Smith-Gill, S.J., Todd, P.E. and Wilson, A.C. (1984) Ann. Rev. Immunol., 2, 67–101.
2. Blake, C.C.F., Mair, G.A., North, A.C.T., Phillip, D. C. and Sarma, V.R. (1967) Proc. Roy. Soc. London B, 167, 365–377.
3. Sarma, R. and Bott, R. (1977) J. Mol. Biol., 113, 555–565.
4. Artymiuk, P.J. and Blake C.C.F. (1981) J. Mol. Biol.152, 737–762.
5. Matthews, B.W., Grutter, M.G., Anderson, W.F. and Remington, S.J. (1981) Nature 290, 334–335.

6. Köhler, G. and Milstein, C. (1975) Nature 256, 495-497.
7. Harper, M., Lema, F., Boulot, G. and Poljak, R.J. (1986) Mol. Immunol., in press.
8. Mariuzza, R.A., Jankovic, D.L.J., Boulot, G., Amit, A.G., Saludjian, P., Le Guern, A., Mazié, J.C. and Poljak, R.R. (1983) J. Mol. Biol. 170, 1055-1058.
9. Jones, T.A. (1978) J. App. Cryst. 11, 268-272.
10. Verhoeyen, M., Berek, C., Jarvis, J.M., Winter, G. (1986) Personal communication.
11. Luzzati, V. (1952) Acta Cryst. 5, 802-808.
12. Amit, A.G., Mariuzza, R.A., Phillips, S.E.V. and Poljak, R.J. (1986) Science 233, 747-753.
13. Amit, A.G., Mariuzza, R.A., Phillips, S.E.V. and Poljak, R.J. (1985) Nature 313, 156-158.
14. Chothia, A.M., Lesk, A.M., Levitt, M., Amit, A.G., Mariuzza, R.A., Phillips, S.E.V. and Poljak, R.J. ((1986) Science 233, 755-757.
15. Kabat, E.A., Wu, T.T., Bilofsky, H., Reid-Miller, M. and Perry, H. (1983) in: Sequences of Proteins of Immunological Interest. U.S. Department of Health and Public Services: 1.
16. Tonegawa, S. (1983) Nature 302, 575-581.
17. Honjo, T. (1983) Ann. Rev. Immunol. 1, 499-528.
18. Lee, B. and Richards, F.M. (1971) J. Mol. Biol. 55,379-400.
19. Chothia, C. and Janin, J. (1975) Nature 256, 705-708.

PREDICTING ANTIGENICITY IN PROTEINS AND THE FUTURE OF SYNTHETIC PEPTIDE VACCINES

M.H.V. Van Regenmortel

Institut de Biologie Moléculaire et Cellulaire, CNRS, 15 rue Descartes

67084 Strasbourg-Cédex France

The apparently unlimited recognition potential of immunoglobulin molecules presents a special problem to investigators interested in biological specificity. It is widely believed that the immune system of higher vertebrates can generate about 10^7 different immunoglobulins (1) and that this is sufficient to provide significant binding capacity for recognizing virtually any possible configuration of organic molecules. The recognition potential of immunoglobulins originates from six loops of hypervariable sequence, comprising a total of about 30 amino acid residues, which generate the two identical functional attachment sites or paratopes of the antibody molecule. The binding specificity of individual antibodies arises from the variations in sequence found in these loops. Three loops are made up of residues of the heavy chain and three from residues of the light chain of the immunoglobulin molecule.

The hypervariable loops, also known as complementary-determining regions (CDR) are located at the tips of the Y-shaped immunoglobulin molecule and interact, to a greater or lesser extent, with the surface of the antigen. In the case of small haptens, the contact area between antigen and antibody is limited (2, 3) while in the case of protein antigens, the contact area may be much more extensive. To date, only one paratope specific for a protein antigen (lysozyme) has been analyzed by X-ray crystallography (4). In this instance 17 residues of the antibody belonging to the six CDRs and 16 residues of lysozyme were found to interact closely. It is, of course, impossible to predict whether the extensive interpenetration of antigen-antibody surfaces observed in the case of the lysozyme-antibody complex will be a general feature of all interactions between paratopes and protein antigens that will be analyzed in the future.

The residues of the antigen that make contact with the paratope constitute an antigenic determinant or epitope of the antigen. The only way to establish if a set of residues in a protein constitutes an epitope is to search for and find an immunoglobulin able to bind it. An epitope is thus a relational concept which needs a complementary paratope for its operational definition (5). However, it should be emphasized that the relational nexus between the two immunological partners is not of an exclusive nature, but that considerable promiscuity is built into the relationship. It has been accepted for a long time that an individual epitope can be recognized by a variety of different paratopes, and it is surprising that the reciprocal situation has become accepted much more slowly (6, 7, 8, 9). The fact that an antibody is polyspecific, i.e. that it is able to react with a variety of more or less closely related antigens, is of crucial significance. Indeed, the cross-reactive binding capacity of antibody molecules is mainly responsible for the fact that we have been able during the last 20 years to acquire some information concerning the nature of protein epitopes. It is, indeed, the ability of anti-protein antibodies to reveal a cross-reaction between an intact protein and peptide fragments derived from it which brought about most of our knowledge concerning protein antigenicity.

Most studies designed to unravel the antigenic structure of proteins have analyzed the antigenic activity of natural or synthetic fragments of the molecule. If a fragment is able to bind to the protein antibodies and therefore able to inhibit the reaction between these antibodies and the intact protein, it is assumed that the fragment represents an epitope of the complete molecule. Such linear peptide fragments are usually referred to as continuous epitopes, to distinguish them from discontinuous epitopes that are made up of residues at the surface of the antigen that are not contiguous in the sequence (10). It should be realized that the label "continuous epitope" attached to a peptide does not mean that the fragment accurately mimics the full structure of the protein epitope, since it cannot be excluded that the peptide is reacting with antibodies induced by a larger discontinuous epitope of the protein. In this case, the linear fragment would represent only a part of a complex discontinuous epitope. In general, the extent of antigenic cross-reactivity between a peptide and the intact protein is very low, which testifies to the limited structural resemblance between the two. Until recently, it was widely assumed that this low cross-reactivity was due to the fact that peptides exist in solution in thousands of conformations, only a few of them being recognized by the antibody. Today, this explanation has fewer adepts since there is growing evidence that peptides in solution have distinct conformational preferences. The most commonly held view states that the vast majority of epitopes are discontinuous and that linear peptides possess a low antigenic cross-reactivity with proteins because they mimic, in a conformationally imperfect way, only part of the epitope structure (11, 12, 5, 13).

In a recent paper (14) it has been argued on purely theoretical grounds, based on the size of the area of contact between a protein and its antibody, that all epitopes are likely to be discontinuous. The validity of such an argument rests on the premise that the number of contact residues of antigen and antibody present in all epitopes will be found to be similar to that observed in the case of the lysozyme-antibody complex (4). It cannot be excluded that only some of the 6 CDRs of certain anti-protein antibodies will, in the future, be found to be implicated in direct contact with residues of the antigen.

Using our current knowledge of the antigenic structure of a few well-studied proteins derived mainly from the approximate location of so-called "continuous epitopes," several authors have attempted to derive algorithms for predicting the position of epitopes from the protein sequence (15, 16, 12, 13, 17). Using relative scales of hydrophilicity, mobility and surface accessibility for each of the 20 amino acids, it is possible to construct plots to predict how these properties are likely to vary along the polypeptide chain. Since local maxima in these plots frequently correspond to the position of continuous epitopes, they give an indication of which peptides should be synthesized for the purpose of mimicking antigenic and immunogenic regions of proteins. It should be emphasized, however, that parameters such as hydrophilicity, accessibility, segmental mobility and chain termination are not independent variables but are interconnected. Any attempt to single out one of these properties as a "primary" explanatory, predictive and causal category for cross-reactive antigenicity between proteins and peptides is bound to fail. Looking for a necessary and sufficient connection in the form of cause and effect is not helpful when trying to explain what is actually a reciprocal interaction between two immunological partners (13).

In recent years there has been considerable interest in the possibility of using synthetic peptides as vaccines against infectious agents and toxins (18, 19, 20, 21, 22, 23). Developing peptides suitable for vaccination is a more difficult task than simply selecting peptides able to induce antibodies that will react with the pathogen or toxin. Since only a small number of the epitopes of a virus are capable of eliciting antibodies that neutralize viral infectivity, it is these neutralization epitopes that have to be mimicked by means of synthetic peptides. Unfortunately, the mechanism of the neutralization process is poorly understood and appears to vary in different systems (24).

Monoclonal antibodies possessing neutralizing activity have been useful in identifying certain surface regions of influenza, polio, foot-and-mouth disease and hepatitis B virus particles that are involved in the neutralization of these viruses. As the atomic structures of influenza, polio and rhino viruses have been established (25, 26, 27) it is becoming easier to localize the position of neutralization epitopes at the surface of these virus particles (28, 29). The central question regarding the development of effective synthetic vaccines is whether it will be necessary to reproduce exactly the tertiary conformation of neutralization epitopes in order to achieve protective immunity by means of synthetic peptides. There is evidence that many viral epitopes are "neotopes," i.e. epitopes that arise from the quaternary structure of the virion and are thus absent in the constituent monomeric viral subunits (30). In poliovirus, for instance, certain antigenic sites are made up of residues from two structural proteins VP1 and VP3. If most neutralization epitopes were of this type, the prospects for new synthetic vaccines would depend on prior knowledge of the three-dimensional structure of the different viruses. However, there is also evidence that protective immunity can be elicited by vaccination with linear peptides of about 20 residues (21,23). It is thus conceivable that the cross-reactive potential of antibodies induced by synthetic peptides will in effect lower the absolute requirement for reproducing exactly the conformational parameters of protein epitopes in synthetic peptides intended for vaccination. The potential of synthetic vaccines against viral, bacterial and parasitic diseases is at present clearly established. As clinical trials are being conducted in several countries, it should soon become clear whether in addition to their purely scientific interest, they also represent a practical and effective means to bring about protective immunity.

REFERENCES

1. Kindt, T.J. and Capra, J.D. (1984) The antibody enigma. Plenum Press.
2. Amzel, L.M., Poljak, R.J., Saul, F., Varga, J.M. and Richards, F.F. (1974) Proc. Natl. Acad. Sci. USA 71, 1427-1430.

3. Segal, D.M., Padlan, E.A., Rudikoff, S., Potter, M. and Davies, D.R. (1974) Proc. Natl. Acad. Sci. USA 71, 4298-4302.
4. Amit, A.G., Mariuzza, R.A., Phillips, S.E.V. and Poljak, R.J. (1986) Science 233, 747-753.
5. Van Regenmortel, M.H.V. (1986) Trends Biochem. Sci. 11, 36-39.
6. Richards, F.F., Konigsberg, W.H., Rosenstein, R.W. and Varga, J.M. (1975) Science 187, 130-137.
7. Cameron, D.J. and Erlanger, B.F. (1977) Nature 268, 763-765.
8. Lane, D. and Koprowksi, H. (1982) Nature 296, 200-202.
9. Al Moudallal, Z., Briand, J.P. and Van Regenmortel, M.H.V. (1982) Embo. J. 1, 1005-1010.
10. Atassi, M.Z. and Smith, J.A. (1978) Immunochemistry 15, 609-610.
11. Benjamin, D.C., Berzofsky, J.A., East, I.J., Gurd, F.R.N., Hannum, C., Leach, S.J., Margoliash, E., Michael, J.G., Miller, A., Prager, E.M., Reichlin, M., Sercarz, E.E., Smith-Gill, S.J., Todd, P.A. and Wilson, A.C. (1984) Annual Rev. Immunol. 2, 67-110.
12. Berzofksy, J.A. (1985) Science 229, 932-940.
13. Van Regenmortel, M.H.V. (1986) in: Protides of the biological fluids (Peeters, H., ed.) pp. 81-86, Vol. 34, Pergamon Press, England.
14. Barlow, J., Edwards, M.S. and Thornton, J.M. (1986) Nature 322, 747-749.
15. Hopp, T.P. and Woods, K.R. (1981) Proc. Natl. Acad. Sci. USA 78, 3824-3828.
16. Karplus, P.A. and Schulz, G.E. (1985) Naturwissenschaften 72, S. 212.
17. Hopp, T.P. (1986) J. Immunol. Methods 88, 1-18.
18. Shinnick, T.M., Sutcliffe, J.G., Green, N. and Lerner, R.A. (1983) Ann. Rev. Microbiol. 37, 425-446.
19. Arnon, R., Shapira, M. and Jacob, C.O. (1983) J. Immunol. Methods 61, 261-273.
20. Chanock, R.M. and Lerner, R.A. (1984) in: Modern approaches to vaccines, Cold Spring Harbor Laboratory.
21. Lerner, R.A., Chanock, R.M. and Brown, F. (1985) in: Vaccines 1985. Molecular and chemical basis of resistance to parasitic, bacterial and viral diseases. Cold Spring Harbor Laboratory.

22. Van Regenmortel, M.H.V. and Neurath, R. (1985) in: Immunochemistry of viruses: the basis for serodiagnosis and vaccines, Elsevier, Amsterdam.
23. Brown, F., Chancok, R.M. and Lerner, R.A. (1986) in: Vaccines 1986. New approaches to immunization. Cold Spring Harbor Laboratory.
24. Mandel, B. (1985) in: Immunochemistry of viruses (Van Regenmortel, M.H.V. and Neurath, A.R., eds.) Elsevier, Amsterdam.
25. Wiley, D.C., Wilson, I.A. and Skehel, J.J. (1981) Nature 289, 373-378.
26. Hogle, J.M., Chow, M. and Filman, D.J. (1985) Science 229, 1358-1365.
27. Rossmann, M.G., Arnold, E., Erickson, J.W., Frankenberger, E.A., Griffith, J.P., Hecht, H.J., Johnson, J.E., Kamer, G., Luo, M., Mosser, A.G., Rueckert, R.R., Sherry, B. and Vriend, G. (1985) Nature 317, 145.
28. Chow, M., Yabrov, R., Bittle, J., Hogle, J. and Baltimore, D. Proc. Natl. Acad. Sci. USA 82, 910-914.
29. Blondel, B., Crainic, R., Fichot, O., Dufraisse, G., Candrea, A., Diamond, D., Girard, M. and Horaud, F. (1986) J. Virol. 57, 81-90.
30. Van Regenmortel, M.H.V. (1982) in: Serology and immunochemistry of plant viruses, p. 302, Academic Press, New York.

FOLDING AND ANTIGENICITY OF THERMOLYSIN FRAGMENTS

Angelo Fontana, Claudio Vita, Daniele Dalzoppo and Irwin M. Chaiken§

Department of Organic Chemistry, Biopolymer Research Centre of CNR, University of Padua, 35131 Padua, Italy.

Fragments of well-characterized globular proteins have been used quite extensively in studies aimed to elucidate the nature of the forces that direct the folding of a polypeptide chain into a specific, biologically active conformation. Usually, it has been found that relatively short fragments of proteins do not possess distinct folded conformations in aqueous solution, often attaining a more or less random conformation of little resemblance to the structure they have in the native parent protein. On the other hand, relatively large protein fragments, especially those corresponding to structural domains in globular proteins, were found to acquire a folded conformation of native-like characteristics and often to retain some of the functional properties of the parent protein (1-5).

---

§National Institutes of Health, NIADDKD, Bethesda, 20205 Maryland, USA.

Abbreviations used: CD, circular dichroism; Tris, tris--hydroxymethyl-aminomethane.

A generally employed method to identify antigenic sites in globular proteins involves the use of relatively short peptides (6-12 residues) prepared either by chemical synthesis or by proteolytic or chemical cleavage of the parent protein antigen (6-9). Usually, these peptides are tested, using inhibition binding assays, for their antigenicity, i.e. binding capability towards polyclonal antibodies prepared using the parent, native protein as an immunogen. These short peptides do show, but not generally, specific binding, but with much lower affinity than the parent intact native protein (10-12). These experiments lead to the belief that proteins contain only a small number of antigenic sites or epitopes situated in highly exposed surface areas of the protein molecule (13).

The use of linear short peptides to delineate the antigenic structure of a protein likely has the intrinsic limitation of allowing identification of "continuous" epitopes or sequential antigenic determinants only. In fact, the protein antigenicity is clearly a surface property and the surface of a globular protein is not dictated by the linear array of a peptide chain, but by the special combination of the side chains of amino acid residues lying quite distant in the polypeptide chain, but brought together in space by the specific folding of the protein. Thus, "discontinuous" epitopes or conformational antigenic determinants likely constitute the majority of the antigenic sites in a globular protein and these last would escape detection if short peptides are used in analyzing the antigenicity of a globular protein (14,15).

From these considerations it follows that the use of quite large fragments of globular proteins would be more useful than short peptides in delineating antigenicity of globular proteins, since these fragments, if folded into a native-like and stable structure, would contain mainly conformational determinants. In this communication, we wish to summarize our studies dealing with folding and antigenic properties of relatively large fragments obtained by chemical and enzymatic fragmentation of thermolysin (16-20). The results obtained allowed us to conclude that the isolated fragments, corresponding to protein domains and subdomains, are able to refold into a stable, native-like structure independently from the rest of the polypeptide chain and that their degree of antigenicity correlates with their conformational stability, i.e. rigidity (20). These results are interpreted in the framework of current thinking of molecular aspects of antigenicity of globular proteins (14,15).

SIZE AND LOCATION OF DOMAINS AND SUBDOMAINS IN THERMOLYSIN

A few years ago, we undertook the study of chemical fragments of thermolysin in order to address the question of the existence, location and properties of protein domains (16-20). The word domain of a protein is currently used to indicate large subassemblies of secondary structure elements which appear tightly packed in the crystal structure of globular proteins (21-25). Visually recognized blocks or folded units are commonly referred to as structural domains to indicate that they are identified by inspection of crystallographically determined protein structures. Wetlaufer (21) emphasized the structural role of domains in globular proteins and proposed that domains could represent intermediates in the folding process of protein molecules. Thus, it is conceivable to suggest that specific segments of an unfolded polypeptide chain first refold to individual domains, which then associate and interact to give the final tertiary structure, much the same as do subunits in oligomeric proteins (18,26).

The major implication of this model of protein folding by a mechanism of modular assembly is that isolated protein fragments corresponding to domains in the intact protein are expected to be able to fold into a native-like structure independently from the rest of the polypeptide chain, resembling in their properties a small molecular weigth globular protein. Indeed, this possibility has been in the meantime tested experimentally in a number of laboratories including our own, using protein fragments obtained by chemical cleavage or limited proteolysis of relatively large proteins (5).

Thermolysin shows a quite peculiar bilobal morphology, with two distinct structural domains of equal size (residue 1-157 and 158-316) and the active site located at the interface between them (27-29). Since location of structural domains by visual inspection of protein models suffers from subjectivity, quantitative methods involving computer algorithms were proposed to identify domains from X-ray elucidated coordinates of a protein structure (30-34). Using automatic methods, it was shown that the visually recognized large domains are decomposable into subdomains, and that these entities are, in turn, further decomposable, and so forth. These algorithms allow a description of globular proteins in terms of a hierarchic architecture, ranging from elements of secondary structure (α-helix and β-strands) to supersecondary structure or folding units, to subdomains and domains, and finally to the whole globular protein (30,31). These analyses of the static crystal structure of proteins were used to propose folding pathways involving the hierarchic condensation of continuous chain modules of the polypeptide chain (folding tree)(30).

When applied to thermolysin, these computer methods consistently permitted the identification of two domains of roughly equal size, subdivided into subdomains. The thermolysin morphology deduced by Wodak and Janin (32) using computer-assisted surface-area calculations is shown in Fig. 1. It is seen that two major domains are identified, 1-151 and 152-316, the NH_2-terminal one being additionally subdivided into two pieces and the COOH-terminal one containing subdomains 212-316 and 239-316.

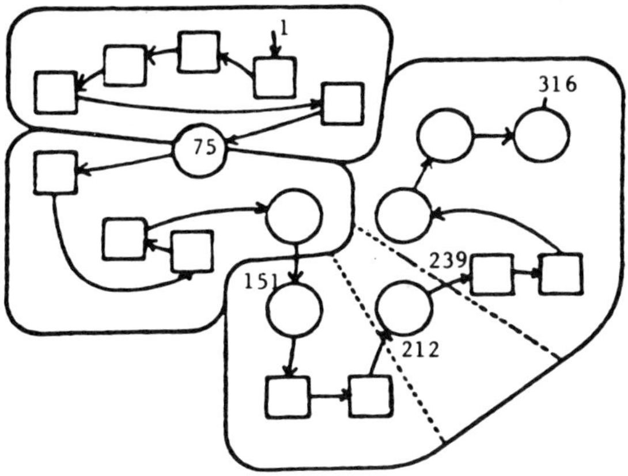

Figure 1. Connectivity diagram of thermolysin. Squares represent β-strands and circles, α-helices. Numbers mark limits of domains and subdomains. Areas enclosed in solid lines indicate domains; those in dashed lines, subdomains (reprinted with permission from ref. 32).

DOMAIN CHARACTERISTICS OF THERMOLYSIN FRAGMENTS

A main effort in our work has been to study fragments obtained by cyanogen bromide cleavage at the level of the two methionine residues in position 120 and 205 of the polypeptide chain of thermolysin (35,36). Under conditions of limited chemical cleavage by cyanogen bromide, it was possible to obtain, besides fragments 1-120, 121-205 and 206-316, also the overlapping fragments 1-205 and 121-316 (37). All these peptides have been isolated and characterized in terms of conformational properties, but the major attention was devoted to COOH-terminal fragments 121-316 (19) and 206-316 (16,38,39). In addition to the cyanogen bromide fragments, also fragment 225(226)-316 was employed in this study. This last fragment was obtained by limited proteolysis of thermolysin by subtilisin, which leads to formation of a "nicked" protein species, called thermolysin S, a partially active derivative of thermolysin composed of fragments 5-224(225) and 225(226)-316 associated in a stable complex (40).

Circular Dichroism. The secondary structure characteristics of fragment 121-316, 206-316 and 225(226)-316 were evaluated by examining their CD spectra in the far-ultraviolet region in aqueous buffer at neutral pH. The CD spectra of all these fragments show two minima of negative ellipticity near 222 and 208 nm with a cross-over at 203 nm, indicating that a substantial portion of the polypeptide backbone is in an α-helical conformation (41). Quantitative analyses of secondary structure from far-ultraviolet CD spectra were carried out using a computer-assisted method (42). The figures thus obtained for the percent α-helical content for the fragments are listed in Table 1, together with the figures expected for a native-like structure of the fragments and calculated from

Table 1. α-Helical content of COOH-terminal fragments of thermolysin. Far-ultraviolet CD spectra were recorded in 20 mM Tris-HCl, 0.1M $CaCl_2$ buffer, pH 7.4. Helicity content was calculated on the basis of the X-ray structure of thermolysin (27-29) or by computer-analysis of CD spectra (42).

Fragment	Number of residues	$[\theta]_{220\ nm}$	Per cent α-helix	
			X-ray	CD
121-316	196	13,000	45	43
206-316	111	13,500	49	47
225(226)-316	81-82	14,800	62	57

the crystallographically determined structure of thermolysin (27-29). Considering the theoretical and experimental limitations of methods of calculation of protein secondary structure from CD spectra (43-46), the figures of α-helix content of the fragments reported in the Table indicate a substantial agreement between the experimentally determined helicity from CD spectra with that expected for a native-like structure of the fragments.

Thermal Unfolding. Protein fragments possessing domain characteristics are expected to behave like small molecular weight proteins, i.e. to show globularity, compactness and cooperativity of unfolding transitions (21). Fig. 2 shows the temperature dependence of the mean residue ellipticity, $[\theta]$, measured at 220 nm, for intact and nicked thermolysin, as well as thermolysin fragments 121-316, 206-316 and 225(226)-316; $[\theta]_{220\ nm}$ is a useful parameter of helical secondary structure in proteins (41). The thermal transitions observed under the conditions used were completely reversible for the three fragments tested, as shown by the fact that the profiles of $[\theta]$ vs. temperature obtained by lowering the temperature, after structural unfolding at high temperature, were essentially superimposable on those obtained initially by heating. Thermal denaturations of both thermolysin and thermolysin S were not reversible, owing largely to the autolytic degradation and precipitation which occurred at high temperature. Inspection of the data shown in Fig. 2 allows evaluation of the relative thermal stabilities of the protein and fragment species tested and gives the following order: 225(226)-316 $\leq$ 206-316 $<$ 121-316 $<$ thermolysin S $<$ thermolysin.

IMMUNOCHEMICAL STUDIES

Even if the quantitative analyses of CD spectra of the isolated fragments indicate that they attain in solution a helical secondary structure related to that exhibited by

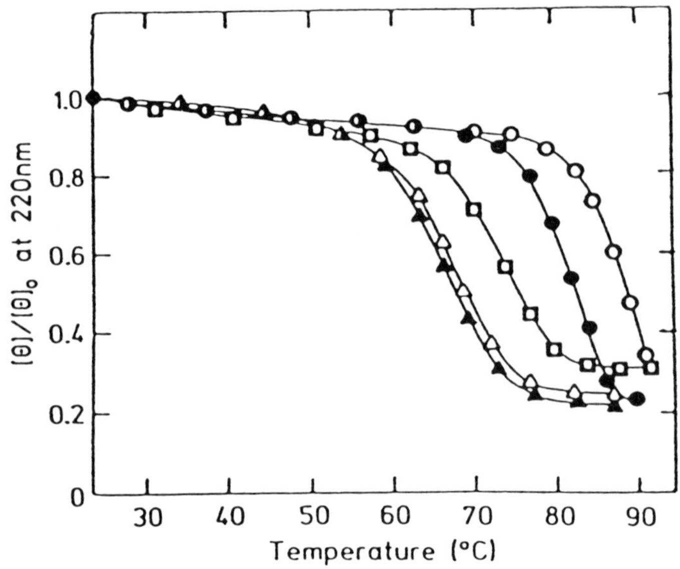

Figure 2. Thermal denaturation profile of thermolysin (○), thermolysin S (●) and fragments 121-316 (□), 206-316 (△) and 225(226)-316 (▲). The mean residue ellipticity at 220 nm was measured as a function of temperature with proteins and protein fragments dissolved in 10 mM morpholinopropane sulfonate buffer, pH 7.8, containing 0.1 M $CaCl_2$. The results are reported as $[\Theta]/[\Theta]_o$ *versus* temperature, where $[\Theta]_o$ is the ellipticity value at 25°C (adapted from ref. 20).

the corresponding chain segments in the native intact protein (Table 1), it cannot be estimated on the basis of these data alone the actual degree of nativeness of their conformation. To this aim, the fragments were studied using antibodies as probes of conformation. This is a well established approach which has been a subject of considerable discussion and experimentation (8,10,47). When a foreign protein molecule is injected into a rat or a rabbit, antibodies towards this protein are mainly produced against conformational determinants which are located in the more exposed regions of the folded polypeptide chain of the globular protein. Consequently, antibodies would contain information related to the detailed features of the antigen (protein) three-dimensional structure.

All the fragments tested (see above) gave precipitin lines in immunodiffusion assays using rabbit anti-thermolysin antibodies (48). These antigen-antibody reactions were shown to be specific and linked to conformational determinants, since denaturated thermolysin fails to give precipitin lines (36). Thus, immunodiffusion assays allowed us to conclude that important structural details (such as loops, corners at the surface of the antigen) are shared between the fragments and the native protein.

A more quantitative use of the immunochemical approach for conformational analysis of the fragments under study was made utilizing affinity-purified antibodies and comparing quantitatively the relative affinities of these antibodies towards isolated fragments and intact as well as nicked thermolysin. To this aim, the rabbit anti-thermolysin serum was sequentially fractionated on columns of Sepharose-thermolysin and then Sepharose--fragment 206-316. In both cases, strongly bound antibodies were eluted from the column with a 3 M guanidine hydrochloride solution and these fractionated antibodies were used for subsequent studies. The product of this double-column fractionation scheme comprises antibodies specific for antigenic determinants located in the sequence 206-316 of native thermolysin (20,49; see also ref. 50, for a discussion on the affinity-purification of antibodies).

Interactions of affinity-purified antibodies with protein fragments and protein species were evaluated by measuring the ability of these antigens to compete with ^{14}C-labeled thermolysin for antibody binding. The results of these competition radioimmunoassay experiments are shown in Fig. 3. Native thermolysin, when added in equimolar ratio with respect to ^{14}C-thermolysin, inhibits 50 % of the

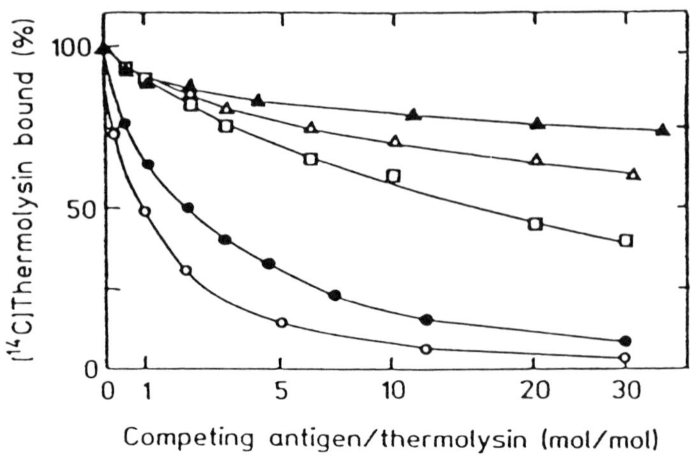

Figure 3. Competition binding profiles of ^{14}C-thermolysin for purified rabbit anti-thermolysin antibodies in the presence of thermolysin (○), thermolysin S (●) and fragments 121-316 (□), 206-316 (△) and 225(226)-316 (▲) as competing antigens (adapted from ref. 20).

binding of labeled protein to the antibodies, indicating that the labeling procedure by carbamoylation with $K^{14}CNO$ did not affect the antigenic integrity of the protein. Inhibition of binding by thermolysin S is quite similar to that observed with native thermolysin, in agreement with the fact that the nicked protein shows conformational properties similar to those of the native one, as judged by the fact that both protein species show the same far-ultraviolet CD spectrum (40). On the other hand, fragments 121-316, 206-316 and 225(226)-316 show distinctly weaker, but nonetheless substantial, competition compared to native thermolysin, with competitive activity increasing with fragment size. In particular, it should be observed that the ranking of the degree of antigenicity (Fig. 3) of the species tested is the same as that of their thermal stability (Fig. 2).

DISCUSSION

Spectroscopic and immunochemical data indicate that the COOH-terminal fragments of thermolysin here investigated are able to refold into a structure resembling that of the corresponding chain segments in the intact, native protein. This native-like structure is attained by fragment 121-316, comprising in full the COOH-terminal "all-α" (51) structural domain 151-316, as well as by the shorter fragments 206-316 and 225(226)-316, which correspond to subdomains of the COOH-terminal domain (cf. Fig. 1). In this context, mention should be made that additional fragmentation of fragment 206-316 by limited proteolysis with subtilisin allowed the isolation of fragment 255-316 in homogeneous form (52). This is the shortest folded fragment of thermolysin yet isolated in our studies, given by a single polypeptide chain of 62 amino acid residues without disulfide bridges or bound ions, able to refold to a highly helical, native-like structure of high thermal stability (Tm 65° C). Of interest, the properties of fragment 255-316 demonstrate that a stable supersecondary structure of thermolysin can be isolated, corresponding to the smaller COOH-terminal subdomain 239-316 of thermolysin identified by a computer algorithm and containing three helices (cf. Fig. 1).

For the thermolysin and fragment species studied here, we have found a significant correlation between stability, as measured by thermal unfolding profiles (Fig. 2), and antigenicity, as measured by degree of competition of native thermolysin to affinity-purified antibodies specific for the sequence 206-316 (Fig. 3). Considering that it has been demonstrated that an inverse correlation exists between flexibility and thermal stability (as measured for example by Tm, temperature of half denaturation) (53-55), the data of Fig. 2 and 3 can be interpreted as indicating a correlation between antigenicity and rigidity of the fragments tested. According to this view, the purified antibodies recognize better the intact and nicked thermolysin than less thermally stable and more flexible fragments. In particular, the shorter and less stable (in terms of Tm) fragment 225(226)-316 is stabilized upon binding to fragment 5-224(225) into a native-like conformation and becomes more antigenic, because conformational fluctuations are "damped" in the tightly packed complex of the thermolysin S molecule.

The above considerations and interpretations of the data shown in Fig. 2 and 3 apparently contrast with the recently proposed correlation between antigenicity and segmental mobility, this last given by the atomic temperature factors (B values or Debye-Waller factors) (56-59). These crystallographically determined B values represent the mean-square displacement of each atom from its average position in the protein crystal and, when plotted against residue number, provide a graphic image of the degree of mobility along the polypeptide chain. The correlation between segmental mobility and antigenicity has been demonstrated by examining the binding to anti-protein antibodies of short (6-8 residues) peptides devoid of any secondary structure. In addition, the same correlation was observed when antibodies were raised against peptide fragments of a well-characterized protein (myohemerythrin) (57). These anti-peptide antibodies recognize best the native protein when the peptides used as immunogens correspond to mobile sites in the protein structure. The conclusion reached from these studies was that mobility of a continuous antigenic determinant may make easier the adjustements of that epitope to the geometry of the antibody combining site (58,59).

However, the conclusions reached by the studies above mentioned are truly valid for the specific experimental approaches employed and cannot be generalized. In fact, as already discussed by Van Regenmortel (60), the use of short peptides as probes of antigenicity would be biased towards the detection of mobile epitopes, since it is expected that these structureless peptides would mimick more easily the mobile sites of that protein. Similarly, antibodies raised towards short peptides would recognize their corresponding intact protein best when the target site in the protein is mobile. Thus, it seems likely that the correlation between mobility and antigenicity is valid for continuous epitopes only. When longer and folded peptides, such as the thermolysin fragments here investigated, are used as probes of antigenicity, the majority of epitopes identified are of discontinuous nature, since only a small part of the folded fragment surface is given by linear stretches of amino acid residues. The conformational details of these discontinuous antigenic sites are made more native-like when the sizes of the fragments increase and additional tightening (rigidification of native-like structure) is achieved, as evidenced by the enhanced thermal stability of the fragment conformation.

In summary, this study provides evidence that rather long and folded peptides can be usefully employed to delineate molecular aspects of protein antigenicity, complementing previous studies based on the use of short, structureless peptides. The results of studies such as those here described, aiming to correlate conformation--stability-antigenicity, could also provide useful guidelines for the design of peptides that will mimick antigenic sites of native proteins sufficiently well to be efficacious as synthetic vaccines.

REFERENCES

1. Anfinsen, C.B. (1973) Science 181, 223-230.
2. Wetlaufer, D.B. and Ristow, S. (1973) Ann. Rev. Biochem. 42, 135-158.

3. Anfinsen, C.B. and Scheraga, H.A. (1975) Adv. Protein Chem. 29, 205-300.
4. Chavez, L.G.Jr. and Scheraga, H.A. (1980) Biochemistry 19, 1005-1012.
5. Wetlaufer, D.B. (1981) Adv. Protein Chem. 34, 61-92.
6. Atassi, M.Z. (1977) in: Immunochemistry of Proteins (Atassi, M.Z., ed.) vol. 2, pp. 77-176, Plenum, New York.
7. Atassi, M.Z. (1978) Immunochemistry 15, 909-936.
8. Crumpton, M.J. (1974) in: The Antigens (Sela, M., ed.) vol. 1, pp. 1-72, Academic Press, New York.
9. Lerner, R.A. (1982) Nature 299, 592-596.
10. Furie, B., Schechter, A.N., Sachs, D.H. and Anfinsen, C.B. (1975) J. Mol. Biol. 92, 497-506.
11. Todd, P.E.E., East, I.J. and Leach, S.J. (1982) Trends Biochem. Sci. 7, 212-216.
12. Leach, S.J.(1983) Biopolymers 22, 425-440.
13. Atassi, M.Z. (1984) Eur. J. Biochem. 145, 1-20.
14. Berzofsky, J.A. (1985) Science 229, 932-940.
15. Benjamin, D.C., Berzofsky, J.A., East, I.J., Gurd, F.R.N., Hannum, C., Leach, S.J., Margoliash, E., Michael, J.G., Miller, A., Prager, F.M., Reichlin, M., Sercarz, E.E., Smith-Gill, S.J., Todd, P.E.E. and Wilson, A. (1984) Ann. Rev. Immunol. 2, 67-101.
16. Fontana, A., Vita, C. and Chaiken, I.M. (1983) Biopolymers 22, 69-78.
17. Vita, C. Dalzoppo, D. and Fontana, A. (1984) Biochemistry 23, 5512-5519.
18. Fontana, A., Vita, C. and Dalzoppo, D. (1985) J. Biosci. 8, 57-66.
19. Vita, C., Dalzoppo, D. and Fontana, A. (1983) Int. J. Peptide Protein Res. 21, 49-56.
20. Vita, C., Fontana, A. and Chaiken, I.M. (1985) Eur. J. Biochem. 151, 191-196.
21. Wetlaufer, D.B. (1973) Proc. Natl. Acad. Sci. USA 70, 697-701.
22. Schulz, G.E. and Schirmer, R.H. (1979) Principles of Protein Structure, pp. 159-160, Springer Verlag, Heidelberg.

23. Rossmann, M.G. and Argos, P. (1981) Ann. Rev. Biochem. 50, 497-532.
24. Janin, J. and Wodak, S.J. (1983) Prog. Biophys. Molec. Biol. 42, 21-78.
25. Liljas, A. and Rossmann, M.G. (1984) Ann. Rev. Biochem. 43, 475-507.
26. Jaenicke, R. (1984) Angew. Chem. Int. Engl. Ed. 23, 395-413.
27. Colman, P.M., Jansonius, J.M. and Matthews, B.W. (1972) J. Mol. Biol. 70, 701-724.
28. Matthews, B.W., Weaver, L.H. and Kester, W.A. (1974) J. Biol. Chem. 249, 8030-8044.
29. Holmes, M.A. and Matthews, B.W. (1982) J. Mol. Biol. 160, 623-639.
30. Crippen, G.M. (1978) J. Mol. Biol. 126, 315-332.
31. Rose, G.D. (1979) J. Mol. Biol. 134, 447-470.
32. Wodak, S.J. and Janin, J. (1981) Biochemistry 20, 6544-6552.
33. Rashin, A.A. (1981) Nature (London) 291, 85-87.
34. Rashin, A.A. (1984) Biochemistry 23, 5518-5519.
35. Titani, K., Hermodson, M.A., Ericsson, L.H., Walsh, K.A. and Neurath, H. (1972) Biochemistry 11, 2427-2435.
36. Vita, C., Fontana, A., Seeman, J.R. and Chaiken, I.M. (1979) Biochemistry 18, 3023-3031.
37. Vita, C., Dalzoppo, D., Patti, S. and Fontana, A. (1984) Int. J. Peptide Protein Res. 24, 104-111.
38. Vita, C. and Fontana, A. (1982) Biochemistry 21, 5196-5202.
39. Dalzoppo, D., Vita, C. and Fontana, A. (1985) Biopolymers 24, 1798-1806.
40. Vita, C., Dalzoppo, D. and Fontana, A. (1985) Biochemistry 24, 1798-1806.
41. Greenfield, N. and Fasman, G.D. (1969) Biochemistry 9, 4108-4115.
42. Siegel, J.B., Steinmetz, W.E. and Long, G.L. (1980) Anal. Biochem. 104, 160-167.
43. Wu, C.-S.C. and Yang, J.T. (1976) Biochemistry 15, 3007-3014.

44. Hennessey, J.P. and Johnson, W.C. Jr. (1981) Biochemistry 15, 688-698.
45. Baker, C.C. and Isemberg, I. (1976) Biochemistry 8, 1947-1951.
46. Chang, C.T., Wu, C.-S.C. and Yang, J.T. (1978) Anal. Biochem. 91, 13-31.
47. Reichlin, M. (1975) Adv. Immunol. 20, 71-123.
48. Outchterlony, O. (1948) Acta Pathol. Microbiol. Scand. 25, 186-191.
49. Vita, C., Fontana, A. and Chaiken, I.M. (1982) Biochemistry 21, 2016-2022.
50. Dean, J. and Schechter, A.N. (1979) J. Biol. Chem. 254, 9185-9193.
51. Levitt, M. and Chothia, C. (1976) Nature (London) 261, 552-558.
52. Dalzoppo, D., Vita, C. and Fontana, A. (1985) J. Mol. Biol. 182, 331-340.
53. Wagner, G. and Wutrich, K. (1978) Nature (London) 275, 247-248.
54. Wutrich, K., Roder, H. and Wagner, G. (1980) in: Protein Folding (Jaenicke, R., ed.) pp. 549-564, Elsevier/North Holland Biomedical Press, Amsterdam-New York.
55. Tsuboi, M., Ohta, S. and Nakanishi, M. (1978) in: Biochemistry of Thermophily (Friedman, S.M., ed.) pp. 251-266, Academic Press, New York.
56. Westhof, E., Altschuh, D., Moras, D., Bloomer, A.C., Mondragon, A., Klug, A. and Van Regenmortel, M.V.H. (1984) Nature (London) 311, 123-126.
57. Tainer, J.A., Getzoff, E.D., Alexander, H., Houghten, R.A., Olson, A.J., Lerner, R.A. and Hendrickson, W.A. (1984) Nature (London) 312, 127-134.
58. Tainer, J.A., Getzoff, E.D., Paterson, Y., Olson, A.J. and Lerner, R.A. (1985) Ann. Rev. Immunol. 3, 501-535.
59. Getzoff, E.D., Tainer, J.A. and Lerner, R.A. (1985) in: Immune Regulation (Feldmann, M. and Mitchinson, N.A., eds.) pp. 243-258, Humana Press, Clifton, N.J.
60. Van Regenmortel, M.H.V. (1986) Trends Biochem. Sci. 11, 36-39.

SYNTHESIS AND BIOLOGICAL ACTIVITY OF THE IMMUNOMODULATORY SITE OF INTERLEUKIN 1

G. Antoni, R. Presentini, F. Perin, A. Tagliabue, P. Ghiara, S. Censini, G. Volpini, L. Villa, L. Nencioni and D. Boraschi

Sclavo Research Center, Siena, Italy

The knowledge of the primary structure of a protein may provide important information on the structure-activity relationship within the molecule, as the zones of the molecule particularly rich in hydrophilic residues are presumably located on its external surface, so that they can interact more easily with other molecules of the environment.

We have adopted this starting point to study the mechanism of action of interleukin 1 (IL-1). This term is used to represent a family of proteins which play many different roles in inflammation and immunity (1). At least two different forms of human IL-1 have so far been identified and sequenced after DNA cloning, with different isoelectric points, called IL-1α and IL-1ß. These two proteins present a low but significant degree of homology (2,3). A murine IL-1 molecule has also been identified (4), with a high degree of homology with human IL-1α.

To clarify the complex biological role of these molecules, we carried out a study aimed at the identification of the possible location of active sites on the protein molecules. We therefore studied the distribution of the hydrophilic residues along the peptide chains, in order to select fragments endowed with high hydrophilicity and therefore more likely to be located on the external part of the molecule.

We performed our studies with two hydrophilicity scales, selected among the many reported in Hopp (5): the classical hydrophilicity scale, proposed by Hopp and Woods for the detection of sequential antigenic determinants (6), and the so-called acrophilicity scale, more recently described by Hopp (7). The latter takes into account the dimensions of the amino acid residues, giving higher values for small residues that are probably found on the external surface of the molecule, due to their low steric hindrance, while cumbersome residues present lower values. The values of the two scales are reported in Table I.

Figure 1 show the hydrophilicity and acrophilicity profiles of human IL-ß. We selected for our studies two peptides with high values of both hydrophilicity

Table I. Hydrophilicity and acrophilicity values of the twenty natural amino acids.

	Hydrophilicity	Acrophilicity
Ala	-0.5	-0.5
Arg	3.0	0.3
Asn	0.2	2.3
Asp	3.0	2.1
Cys	-1.0	-2.6
Gln	0.2	-0.2
Glu	3.0	0.5
Gly	0.0	3.0
His	-0.5	-.4
Ile	-1.8	-2.5
Leu	-1.8	-2.5
Lys	3.0	1.4
Met	1.3	-1.8
Phe	-2.5	-2.7
Pro	0.0	2.6
Ser	0.3	1.8
Thr	-0.4	-0.1
Trp	-3.4	-3.0
Tyr	-2.3	-0.2
Val	-1.5	-1.7

and acrophilicity, corresponding to the two peaks indicated by the arrows in the Figure (8). The sequence of the fragment which includes the two peaks is reported in the bottom of the Figure, together with the corresponding fragment of murine IL-1, superimposed onto that of human IL-1 when the two sequences are suitably aligned (9).

The peptides within boxes, i.e. fragments 163-171 and 189-196 of human IL-1ß and 175-183 and 199-209 of murine IL-1, were synthesized by the solid phase method (10), with a chloromethylated polystyrene-1% divinylbenzene copolymer, using N-αBoc protected amino acids. At the end of the synthesis the peptides were

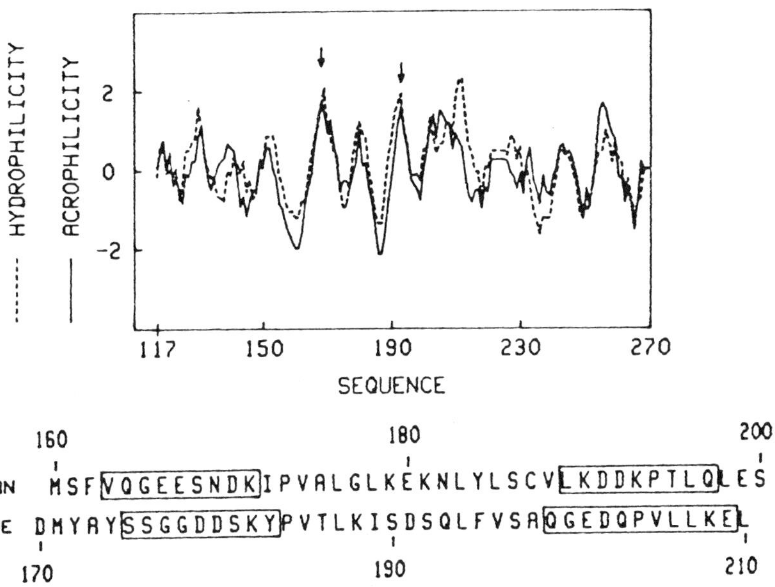

Figure 1. Hydrophilicity (........) and acrophilicity (————) profiles of human IL-1ß. Each point of the lines represents the average hydrophilicity or acrophilicity, calculated on six successive residues. Adapted from reference 8, with permission of the publishers.

de-protected and detached from the resin with HF, and purified by preparative high performance liquid chromatography (8).

We first examined the IL-1 activity of these peptides in the classical thymocyte stimulation test. When added to cultures of PHA-stimulated C3H/HeJ thymocytes, peptide 163-171 showed significant activity, comparable at the highest doses to that of the entire molecule (Figure 2). On the other hand, peptide 189-197 and the two murine peptides did not show any appreciable

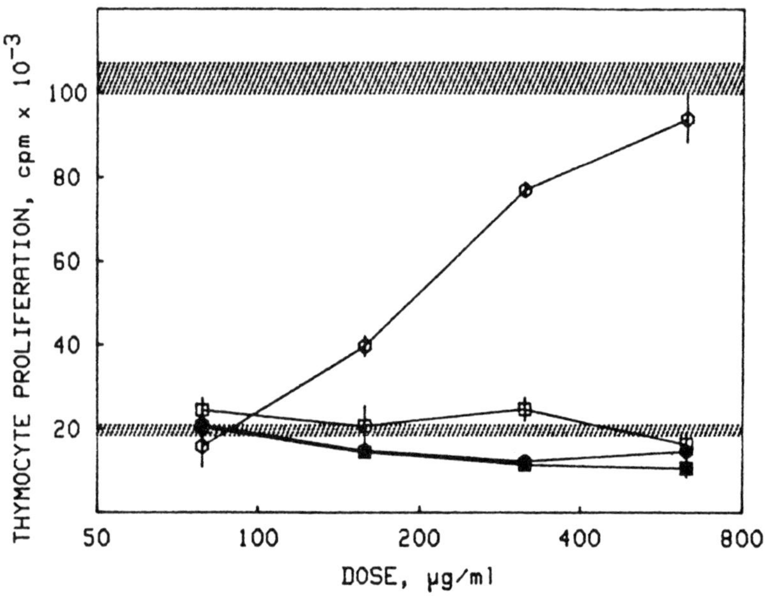

Figure 2. Proliferation of PHA-stimulated C3H/HeJ thymocytes in response to synthetic peptides of IL-1. o human 163-171; ● human 189-197; □ murine 174-182; ■ murine 199-208. Bottom shaded area: control cultures (thymocytes + PHA), mean ± standard error. Top shaded area: thymocyte proliferation in the presence of 15 U/ml of human IL-1 (Genzyme), mean ± standard error. Reproduced from reference 8, with permission of the publishers.

activity. The activity of the 163-171 peptide was not due to IL-2-like activity. In fact, as shown in Table II, the peptide did not stimulate the proliferation of IL 2-dependent CTLL-2 cells. However, as has been demonstrated for IL-1 (11), the peptide induces the production of IL-2 in Concanavalin A-stimulated spleen cells (Table II).

The peptide also showed immunostimulatory activity _in vivo_. In fact, as shown in Figure 3, when the peptide was injected into C3H/HeJ mice at the same time as sheep red blood cells, it was able to enhance the response to the immunizing antigen, measured as plaque forming cells/spleen, in a dose-dependent fashion.

However, the peptide does not appear to be involved in the inflammatory activity of IL-1. In fact, unlike the entire IL-1 molecule, it proved completely unable to stimulate synthesis of prostaglandin E_2 in human dermal fibroblasts _in vitro_ and proved to be devoid

Table II. IL-2-inducing capacity of the peptide fragment 163-171 of human IL-1ß (adapted from reference 8, with permission of the publishers).

Additions to CTLL-2		CTLL-2 proliferation (Stimulation index ± SEM)	IL-2 (U/ml)
nothing		1.00 ± 0.12	0.00
hu IL-1	30 U/ml	0.93 ± 0.02	0.00
163-171	0.84 mg/ml	0.96 ± 0.11	0.00
	0.42 mg/ml	1.03 ± 0.48	0.00
	0.21 mg/ml	1.18 ± 0.23	0.00
ConA-Spleen supernatants :			
+ nothing		6.00 ± 0.20	0.01
+ hu IL-1	30 U/ml	15.27 ± 0.88	0.15
+ 163-171	0.84 mg/ml	149.40 ± 4.65	2.77
	0.63 mg/ml	122.80 ± 11.43	2.25
	0.42 mg/ml	72.07 ± 3.43	1.09

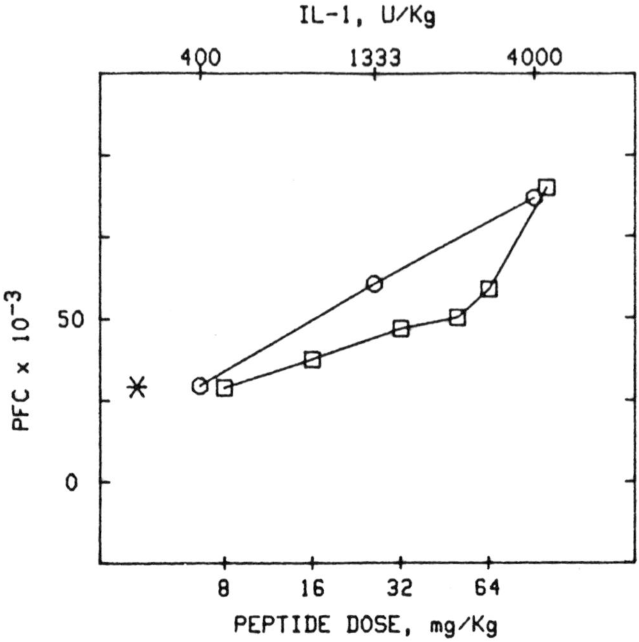

Figure 3. <u>In</u> <u>vivo</u> immune response to sheep red blood cells. Mice were inoculated with sheep red blood cells alone (*) or together with hu IL-1 (o) or the synthetic peptide 163-171 (□). The plaque-forming cell (PFC) response in the spleen was measured after 4 days.

of any pyrogenic activity when inoculated in C3H/HeJ mice (8).

Our results suggest that the immunomodulatory and inflammatory effects of the IL-1 depend on different portions of the molecule, the peptide 163-171 representing a fragment responsible for immunomodulatory effects, devoid of the inflammatory properties characteristic of the entire molecule.

The lack of inflammatory effects in this biologically active peptide makes it particularly interesting for its possible therapeutic applications, besides the theoretical interest for a better understanding of the structure-activity relationship of the IL-1 molecule.

It is interesting to note the relatively high concentrations of the synthetic peptide needed to observe its biological effects, when compared to the very high specific activity of the entire molecule. In actual fact, a solution of short peptide is generally composed of a mixture of many different conformations in equilibrium with each other, while the biological activity is possibly exerted by the peptide only when it is in the conformation peculiar to the native protein. Furthermore, a cooperative contribution of other parts of the molecule might be necessary for a complete manifestation of the biological activity. In any case, as the peptide appears devoid of inflammatory effects, it could be used safely also in large amounts *in vivo* for possible therapeutic applications, thus compensating for its low specific activity.

REFERENCES

1. Dinarello, C.A. (1984) Rev. Infect. Dis. 6, 51-95.
2. Auron, P.E., Webb, A.C., Rosenwasser, L.J., Mucci, S.F., Rich, A., Wolff, S.M. and Dinarello, C.A. (1984) Proc. Natl. Acad. Sci. USA 81, 7907-7911.
3. March, C.J., Mosley, B., Larsen, A., Cerretti, D.P., Braedt, G., Price, V., Gillis, S., Henney, C.S. Kronheim, S.R., Grabstein, K., Conlon, P.J., Hopp, T.P. and Cosman, D. (1985) Nature 315, 641-647.
4. Lomedico, P.T., Gubler, U., Hellmann, C.P., Dukovich, M., Giri, J.G., Pan Y.,C.E., Collier, K., Semionow, R., Chua, A.O. and Mizel, S.B. (1984) Nature 312, 458-462.
5. Hopp, T.P. (1986) J. Immunol. Meth. 88, 1-18.
6. Hopp, T.P. and Woods, K.R. (1981) Proc. Natl. Acad. Sci. USA 78, 3824-3828.
7. Hopp, T.P. (1984) in: Synthetic Antigens (Neri, P. and Piazzi, S.E., eds.) 47-60, Annali Sclavo, Siena.
8. Antoni, G., Presentini, R., Perin, F., Tagliabue, A., Ghiara, P., Censini, S., Volpini, G., Villa, L. and Boraschi, D. (1986) J. Immunol. 137, 3201-3204.
9. McLachlan, A.D. (1971) J. Mol. Biol. 61, 409-424.

10. Erickson, B.W. and Merrifield, R.B. (1976) in: The Proteins (Neurath, H. and Hill, R.L., eds.) 3rd ed., Vol. II, pp. 255-527, Academic Press, New York.
11. Smith, K.A., Lachman, L.B., Oppenheim, J.J. and Favata, M.F. (1980) J.Exp. Med. 151, 1551-1556.

THE 208-222 REGION OF THE PEPLOMER GLYCOPROTEIN AS A PUTATIVE BINDING SITE OF RABIES VIRUS WITH THE NICOTINIC ACETYLCHOLINE RECEPTOR

P.Neri*°,L.Bracci[+],A.Di Tommaso*,L.Lozzi*, M.Rustici*,A.Santucci*,P.Soldani[+],S.Petreni*, N.Niccolai°,P.Mascagni§,G.Siligardi§ and W.A.Gibbons§.

*CRISMA;°Dipartimento di Chimica;[+]Dipartimento Biologia Evolutiva,Università di Siena
§School of Pharmacy, University of London

There is increasing evidence that viruses utilize receptors for physiological ligands in order to infect their host cells. For instance, it has been shown that vaccinia virus binds to the receptors of epidermal growth factor (1), Eppstein-Barr virus to the receptor for the complement on B-lymphocytes (2), and those of HTLV-III and rabies to the T4 antigen of T lymphocytes (3) and acetylcholine receptor (4), respectively. Since the acetylcholine receptor is the target organ of many other non-physiological ligands as, for instance, the neurotoxins of snake venoms, the structure of which is well-known, we decided to compare the amino acid sequence of the glycoprotein peplomers of rabies virus (RGV) with that of snake neurotoxins (NTX) in order to detect homologies, which in turn can lead to the identification of the binding site of the viral protein with the receptor.

We found that regions 208-218 of the rabies virus glycoprotein and 30-40 of King Cobra neuro toxin exhibit a high degree of homology, with 8 coincidences out of

11 residues and 1 high acceptable replacement between Asn_{213} and Ser_{35}. Moreover, the region considered is believed to contain the active site of neurotoxin (5).

We also noticed that the degree of homology around Arg_{34}, which is thought to mimic the cationic moiety of the acetylcholine molecule (6), is maximal for the proteins compared.

We then decided to further investigate whether the homology in the primary structure corresponds to similarities in the conformation. We synthesized a tetradecapeptide corresponding to region 209-222 (RVG/209-222) of the rabies virus glycoprotein and studied its conformation through physical-chemical methods, as well as studying the cross-reactivity of its antibodies with the protein from which the peptide sequence was derived.

Our results indicate the existence of an equilibrium of conformers for the peptide in solution (one of which contains a reverse turn centered around the critical Arg residue), a type of structure coinciding with that of the active site of the snake neurotoxins.

We also found that the peptide, when conjugated with key-hole limpet haemocyanin (KLH), is able to elicit antibodies which cross-react with rabies virus, thus suggesting its potential use as a synthetic vaccine.

MATERIALS AND METHODS

Computer studies. The data base of the National Biomedical Research Foundation of Georgetown University (P.I.R.) was used as the source of amino acid sequences. Hydrophilicity profiles and secondary structure predictions were determined by the algorithms of Hopp and Woods (7) and Chou and Fasman (8) respectively.

Synthesis of RVG/209-222. A Model 430A automatic solid-phase synthesizer (Applied Biosystems, Foster City, CA) was used, employing t-BOC chemistry. A coupling percentage of 99.99 was obtained at each step. The peptide was purified by passing it over a 2.5 x 90 cm Sephadex G-10 (Pharmacia, Uppsala, Sweden) column in 0.5 M acetic acid. The HPLC profile on a C18 reverse-phase column showed a single peak, when eluted with a linear gradient of 0.1% trifluoroacetic acid/methanol in 30 min. Amino acid composition and sequence were determined by an automatic analyzer (Kontron, Chromacon 50, Zurich, Switzerland) and a gas-phase microsequencer (Applied Biosystems, Model 470A, Foster City, CA) respectively, and found in agreement with the expected structure.

Antibodies. RVG/209-222 was conjugated with keyhole limpet haemocyanin (KLH) with glutaraldehyde according to Avrameas and Ternynck (9) and 50 mcg/0.1 ml of complete Freund adjuvant were injected into the foot pad of BALB/c mice. Two weeks later the animals were boosted with 50 mcg/0.1 ml of incomplete Freund adjuvant. Three days later, popliteal lymph nodes were removed and used for fusion to obtain monoclonal antibodies according to Cianfriglia et al. (10). Families were screened by enzyme immunoassay (EIA) and a competitive radioassay for acetylcholine receptor (RRA).

EIA. 96-well plates for tissue culture (Costar, Cambridge, MA) were coated overnight at 31°C with 50 mcl/well of a suspension of rabies virus (CVS strain, kindly supplied by Sclavo, Siena, Italy) purified by 2.3 M $(NH_4)_2SO_4$ precipitation and saturated with 3% bovine serum albumin (BSA). Plates coated with RVG/209-222 were prepared using its complex with BSA obtained through a different coupling method (11). Binding was detected using goat anti-mouse IgG conjugated with horse radish peroxidase (Bio-Rad, Richmond, CA) and a Bio-Rad Model 2550 EIA reader.

RRA. The inhibition produced by monoclonals upon binding of α-Bungarotoxin (BTX) (Sigma, St. Louis, MO) to the nicotinic acetylcholine receptor (AchR) was estimated by incubating family supernatant with various dilutions of AchR in buffer A (0.01 M sodium phosphate pH 7.5 containing 0.015 M EDTA, 0.015 M EGTA and 0.01 M iodacetamide) and a fixed amount (10^5 c.p.m.) of I^{125}-labeled toxin for 4 hours at room temperature. A Triton extract of electric organ membranes from Torpedo marmorata according to Lindstrom et al. (12) was used as ligand. The receptor-toxin complex was precipitated adding 0.5 ml of 0.14% bovine IgG in buffer A and 0.5 ml of 18% polyethylene glycol 6000 in the same buffer, incubated 30 min. at room temperature, centrifuged for 30 min.

at 4°C at 2000xg and counted in a Packard Minimaxi 5000.

NMR and CD measurements. NMR spectra were recorded with a Varian XL-300 spectrometer operating at 75.43 MHz for the 13-C nucleus. Assignment of the protonated carbon signals was based on chemical shift considerations and the analysis of the chemical shift correlated 2D spectrum. Spin-lattice relaxation rates were measured by using the standard inversion recovery pulse sequence. All the experiments were run at 25°C and pH 2.7.

The circular dichroism spectra were recorded on a JASCO J-40CS spectrometer with a JOUAN attachment for low temperature measurements in the +60/-135°C range.

RESULTS

Sequence studies. Comparison among the primary structure of the elapid "long" neurotoxins and that of rabies virus peplomer glycoprotein produced a best fit among the residues of regions 30-45 and 208-223 respectively, as reported in Table I.

The following considerations can be made: a) a remarkable number of coincidences (9 out of 16 residues compared) exist between RVG and region 30-45 of NTX, which is considered to be the active region of elapid neurotoxins (the so-called "toxic loop") (5). b) the best fit is with the region around R_{37} and G_{38} of NTX, which are at position n+3 and n+4 of a reverse turn (5). This location allows enough freedom to the side chain of R_{37} to make a bridge with the "distant" D_{31} and so mimic the structure of the acetylcholine, with the positive charge of the guanido group accounting for the cationic moiety of the acetylcholine (6). c) despite the lack in RVG of the short loop existing in the NTX between C_{30} and C_{34}, which brings the side chain of D_{31} close to that of R_{37}, the alignment of the sequence of RVG with that of the short NTX (in which the short loop is also absent) shows the coincidence of N_{213} with the critical D_{31} , a replacement which should still allow the bridge with R_{37}. d) the major differences, with the remarkable exception already discussed at position 34 of NTX and 212 of RVG, occur at residues where the

Table I. Homologies in the primary structure of long and short elapid neurotoxins (NTX) and rabies virus glycoprotein (RVG).

1	1	6	4	1	6	7	1	1	1	4	4	4	1	1	1	VARIABILITY (a
		A			S	I										
		G			T	M				V	E					replacements(b
		D	W		G	Q				I	I	E				
C	D	N	F	C	A	S	R	G	K	R	V	D	L	G	C	long NTX(30-45)(c
*	*		*			*	*	*	*	*						
C	D	I	F	T	N	S	R	G	K	R	A	S	K	G	S	RVG (208-223)(d
*	*					*	*	*	*	*				*		
S	D	-	-	-	-	H	R	G	T	R	I	E	R	G	C	short NTX (30-45)(e
R						F			S	I	T					
									Y							replacements (f
4	1	-	-	-	-	2	1	1	4	3	3	1	1	1	1	VARIABILITY (g

a) Variability calculated according to Kabat et al. (13) on 12 sequences of long elapid NTX. b) Replacements in the sequences of long elapid NTX. c) Sequence of Forest Cobra NTX. d) Sequence of rabies virus peplomer glycoprotein CVS strain. e) Sequence of Sea Snake NTX and f) its possible replacements. g) As a), but calculated on 12 short NTX.

variability calculated for the various NTX is highest.

Moreover, we compared the sequence of RVG with that of the corresponding glycoprotein of Vesicular Stomatitis Virus, a virus which shares many morphological similarities with that of rabies, but not the neurological effects (14). We found that the critical R_{215} is replaced by a negatively charged aspartyl residue, despite the high degreee of homology existing between these proteins.

These considerations strongly suggest that the 208/223 region of RVG closely resembles the "toxic loop" of

NTX, as far as the primary structure is concerned.

This view is also supported by the study of the hydrophilicity profile of RVG, which indicates that the 208-223 region corresponds to a peak of hydrophilicity, suggesting its exposure on the glycoprotein surface (Figure 1 a).

Moreover, the prediction of the secondary structure for RVG (Figure 1 b) indicates a high potential for the existence of a reverse turn with R215 and G216 at positions n+2 and n+3 in the same region of NTX (Figure 2).

Conformation Studies. The solution structure of the 209-222 fragment of the rabies virus glycoprotein has been carried out by a combined analysis of circular dichroism (CD) and nuclear magnetic resonance (NMR) spectroscopic data with the general experimental approach described below:

1) CD studies in water solution at different temperatures in a range of chemical stability and biological significance (typically 0-60°C) for investigating the presence of structural equilibria.
2) Solvent and temperature dependence studies of CD spectra to characterize the single components of the structural equilibrium.
3) Proton and 13-C NMR studies to detect spectral features which can be related to ordered species in solution.
4) Proton-proton Overhauser effect (NOE) measurements for observing short distance intra- or inter-residue dipolar interactions, i.e. through space proximities.
5) 13-C spin-lattice relaxation studies of protonated carbon atoms for obtaining information on molecular motions within the peptide backbone and side chains.

CD spectra of water solutions of the RVG/209-222 peptide suggest that in this solvent it is involved in a structural equilibrium. CD spectra of the peptide in ethanediol and trifluoroethanol contain spectral features which are present in the CD spectrum obtained in water at room temperature (Figure 3). The use of these cryogenic

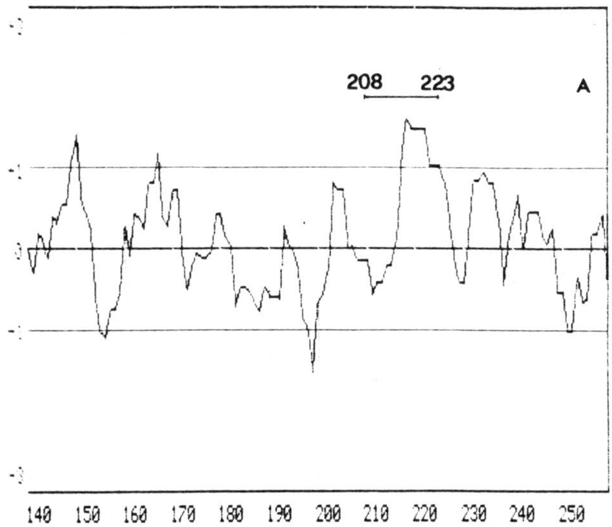

Figure 1.A) Hydrophilicity profile of Rabies Virus Glycoprotein

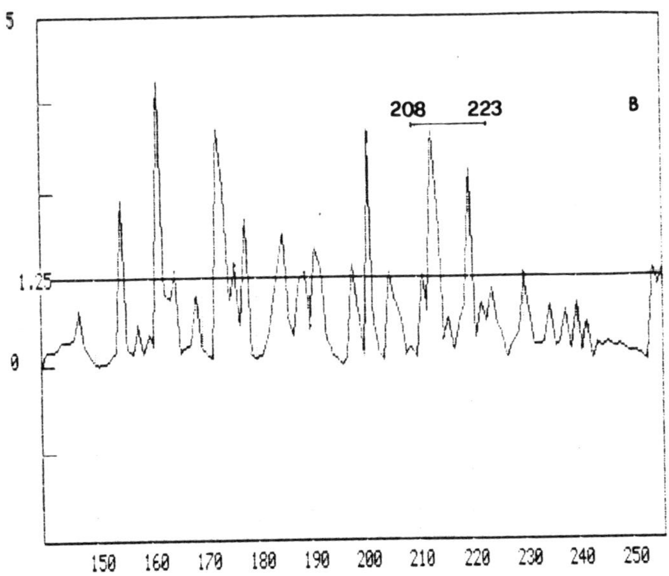

Figure 1.B) Reverse turn propensity of Rabies Virus Glycoprotein

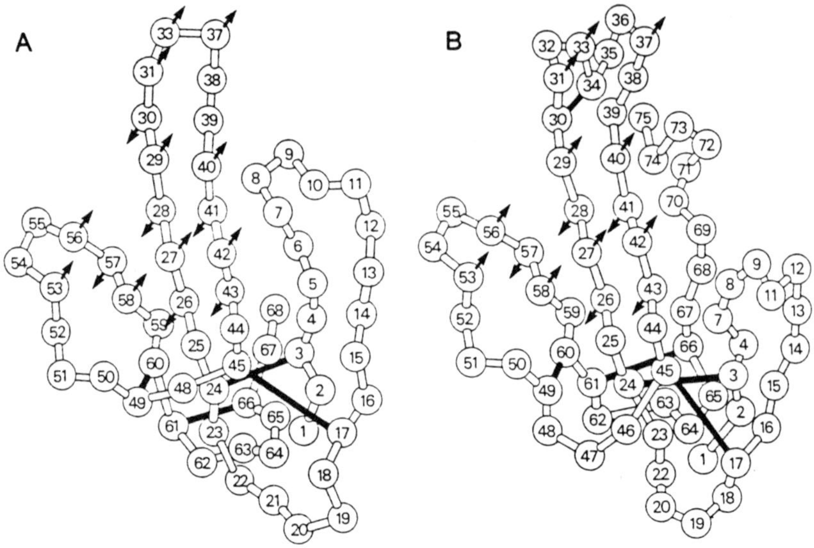

Figure 2. Main chain folding of a short neurotoxin (A) and a long neurotoxin (B), according to the results of X-ray crystallography (Kimball et al., 1979; Walkinshaw et al., 1980). Amino acid residues are numbered taking into account deletions occurring in the short neurotoxin between positions 31 and 38 in order to align homologous residues. The reverse turn having R37 at position n+2 is considered to be involved in receptor binding. According to current models, the side chains of R37 and D31 would pair and mimick acetylcholine, with the guanido group accounting for the cationic moiety of acetylcholine. Distant D31 is brought in contact with R37 by the small loop C30-C34 in the long toxin and as a consequence of four deletions in the short one.
(Reprinted with permission from: Endo et al., Biochemistry 25, 396, 1986. Copyright 1986, American Chemical Society).

solvents allows very low temperature studies so that a fast exchange process can also be frozen into a single conformational state. Such a freezing of the conformational equilibrium was observed for the ethandiol solution, since the CD spectrum does not change at temperatures below -110°C. In these conditions the CD spectrum, with its positive maximum at 210 nm and the two negative maxima at 190 and 200 nm, is characteristic of a reverse turn (15,16). A fitting of the experimental results obtained at different temperatures and solvents with the CD spectrum obtained in water at room temperature indicates that in the latter solution a conformational equilibrium among different folded structures exists.

The circular dichroism results are of primary importance for the interpretation of NMR data. In fact, the lack of proton-proton inter-residue NOEs observed for the peptide in water could be interpreted in terms of the presence in solution of extended random structures. The observation of chemical shift non-equivalence for homologous residue signals in the 1-H and 13-C spectra and the above mentioned CD results suggest that the absence of detectable NOEs is due rather to a distribution of internuclear distances which quenches any structural dipolar coupling.

Spin-lattice relaxation studies of protonated 13-C nuclei can yield powerful information also in the presence of a conformational exchange, since the experimental relaxation rates can be related to average mobilities of the different molecular sites of the peptide.

As shown in Figure 4, alpha carbons of the RVG peptide relax differently, indicating that different molecular motions are effective along the backbone. These relaxation data are consistent with a rather stable folded structure, a beta turn, centered at the S6 and R7 residues. This structure exists together with other more labile conformations where all the residues, excluding G14, are involved via hydrophobic interactions and/or hydrogen bonding.

Binding Sites. We were unable to observe any displacement of I^{125}-labeled BTX by RVG/209-222 in a competitive radioassay for the acetylcholine receptor (experiments not shown).

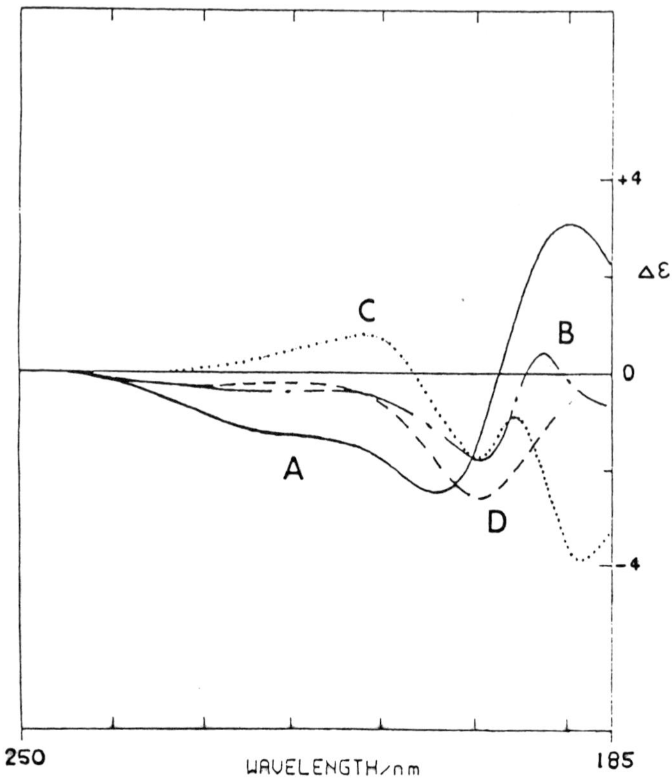

Figure 3. The CD spectra of RVG/209-222: (A,——) in Trifluoroethanol (TFE)/H_2O (3:1) at 24°C at 0.098 mg/ml and path length =0.1cm, (B,-.-.-.) calculated CD spectrum assuming a conformational equilibrium between 50% spectrum C and 50% spectrum A, (C and D) in Ethanediol /H_2O (2:1) v/v at 0.395 mg/ml at path length = 0.02 cm at(....) -110°C and (----)+24°C.

This can be explained by the fact that only a small percentage of the peptide assumes the pleated structure approaching that of the toxic loop, as shown in the previous section concerning the solution conformation of RVG/209-222.

It is then quite possible that by stabilizing such a structure (e.g. by cyclization) the binding becomes possible, provided that more stringent structural requirements do not exist, insufficiently met by the fragment of the sequence considered.

On the other hand, families of monoclonal antibodies obtained against RVG/209-222, which specifically recognize the rabies virus (but not the peptide and viceversa) were able to inhibit the binding of BTX to AchR (Table II).

This finding can be interpreted by supposing that the 209-222 sequence behaves as a conformational epitope

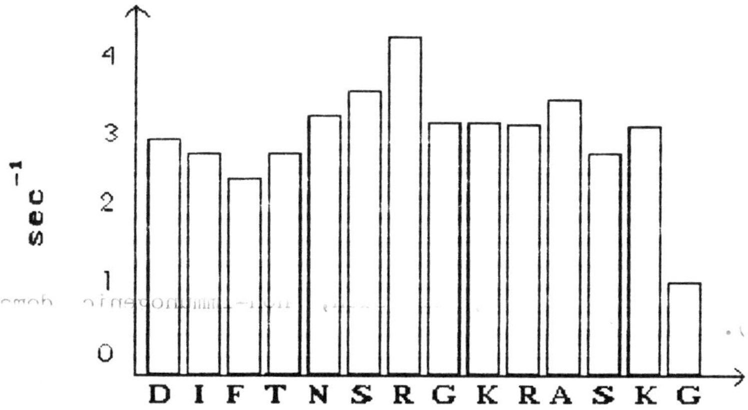

Figure 4. Spin lattice relaxation rates of the αcarbons of RVG/209-222. Values are reported versus peptide sequence.

Table II. Binding of families of anti-RVG/209-222 monoclonal antibodies to various ligands.

FAMILY #	EIA (a) (RVG/209-222)	EIA (a) (Rabies Virus)	RRA (b) (BTX)
13	-	+++	24.3
29	+	-	n.d.
36	-	+	9.0

(a) +: 0 < O.D. < 0.05; ++: 0.05 < O.D. < 0.1; +++: 0.1 < O.D. < 0.15.
(b) % of binding inhibition of BTX to 10^{-7} M AchR Triton extract.

<u>in the structure of the glycoprotein</u> but not <u>in the peptide</u>, due to flexibility of the latter and that the antibodies, which recognize the peptide, are directed against random conformers.

They also suggest that antibodies recognize similar structures in rabies virus and in BTX.

Since the region of RVG, which contains the 209-222 sequence, is considered non-immunogenic (17), the fact that the peptide was able to elicit antibodies against the virus has two meanings. First, it proves once again that antibodies against any region of a protein can be obtained through the technique of synthetic immunogens. Second, it is possible that the 209-222 sequence contains a functional site of the virus corresponding to an invariant and, as such, non-immunogenic domain (18).

CONCLUSIONS

The results of these studies allow us to draw the following conclusions:

1) The region included between C208 and G222 of

peplomer glycoproteins of the rabies virus probably contains the binding site for the nicotinic acetylcholine receptor the virus uses to concentrate in order to attack nerves. This statement is supported by homologies existing in the glycoprotein and neurotoxin sequences especially at those residues, which are critical for the binding to the receptor and by the fact that RVG/209-222 is able to produce antibodies which cross-react with BTX.

2) RVG/209-222 is an interesting candidate in order to develop a polyvalent vaccine against rabies, since the immune response induced by the synthetic peptide is likely to be directed toward a site, the function of which is involved in the infection mechanism of the virus. Thus it can be presumed that the site structure does not vary among the virus strains, and for this reason the relative antibodies could display a wide cross-reactivity. For this same reason it is not surprising that the region including the presumed binding site is considered poorly immunogenic as a part of the whole antigen (17). Thus the use of a synthetic vaccine represents the sole possibility of eliciting an immune response toward such a critical site.

Our studies also proved that the peptide is involved in a conformational equilibrium. Improvement in the immunological activity could be expected by stabilizing the active structure through modification of RVG/209-222 (e.g. cyclization). In this respect, a variety of analogs are currently being investigated in our laboratories.

Acknowledgements. This work was supported by a grant from the Consiglio Nazionale delle Ricerche, Progetto Finalizzato Chimica Fine e Secondaria.

REFERENCES

1. Eppstein, D.A., Marsh, Y.V., Schreiber, A.B., Newman, S.R., Todaro, G.J. and Nestor, J.J. Jr. (1985) Nature 318, 663-665.
2. Fingeroth, J.D. et al. (1984) Proc. Natl. Acad. Sci. USA 81, 4510-4514.
3. Klatzmann, et al.(1984) Nature 312, 767-768.

4. Lentz, T.L., Burrage, T.G., Smith, A.L., Crick, J. and Tignor, G.H. (1982) Science 215, 182-184.
5. Malcolm, D., Walkinshaw, W.S. and Maelicke, A. (1980) Proc. Natl. Acad. Sci. USA 77, 2400-2404.
6. Tsernoglou, D., Petsko, G.A. and Hudson, R.A. (1978) Mol. Pharm. 14, 710.
7. Hopp, T.P. and Woods, K.R. (1981) Proc. Natl. Acad. Sci. USA 78, 3824-3828.
8. Chou, P.Y. and Fasman, G.D. (1978) Adv. Enzymol. 47, 48-148.
9. Avrameas, S. and Ternynck, T. (1969) Immunochemistry 6, 53.
10. Cianfriglia, M., Mariani, M. Armellini, D., Masson, E.A., Lafata, M., Presentini, R. and Antoni, G. (1986) Meth. Enzym. 121, 193.
11. Han, K.K., Richard, C. and Delacourte, A. (1984) Int. J. Biochem. 2, 129.
12. Lindstrom, J., Einarson, B. and Tzartos, S. (1981) Meth. Enzymol. (1981) 74, 432-460.
13. Kabat, E.A., Wu, T.T. and Bilofsky, H. (1976) in: Variable Regions of Immunoglobulin Chains, Bolt Beranck and Newman, Cambridge, MA.
14. Rose, J.K., Doolittle, R.F., Anilionis, A., Curtis, P.J. and Wunner, W.H. (1982) J. of Virology 43, 361-364.
15. Woody, R.W. (1974) in: Peptides, Polypeptides and Proteins, Wiley, New York, pp. 338-358.
16. Gierasch, C.M., Debev, C.M., Madison, V., Niv, C.H. and Blout, E.R. (1981) Biochemistry 20, 4730-4738.
17. Roderick, I. and MacFarlan, et al. (1984) J. Immunol. 133, 2748-2752.
18. Kieber Emmons, T. and Kohle, R.H. (1986) Proc. Natl. Acad. Sci. USA 83, 2521-2525.

POST SCRIPTUM

When this paper was written the Authors were not aware that Lentz et al. had earlier performed sequence comparisons like those presented here and had reached similar conclusions (Lentz, T.L., Wilson, P.T., Hawrot, E. and Speicher, D.W. (1984) Science, 226, 847-848).

HETEROGENEITY AND SPECIFICITY OF MURINE CELL RESPONSE TO SYNTHETIC PEPTIDE $(NANP)_n$ OF *P. FALCIPARUM* CIRCUMSPOROZOITE PROTEIN

G. Corradin*, G. Del Giudice°, A.R. Togna^, A. Verdini^, F. Bonelli^, A. Pessi^, P.H. Lambert°, and H.D. Engers°

*Institute of Biochemistry, Univ. of Lausanne 1066 Epalinges, °WHO Immunology Research Training Center, Univ. of Geneva, 1211 Geneva 4, Switzerland; ^Eniricerche, Polypeptide Synthesis Department, 00015 Monterotondo, Italy

The major antigenic protein of the *P. falciparum* circumsporozoite (CS) is comprised of about 400 amino acids (1,2). The main feature of this protein is the presence of repeats consisting of four amino acids, ASN-ALA-ASN-PRO (NANP; Figure 1).

The number of repeats varies depending upon *P. falciparum* isolates studied. In particular, it was found that a Brazilian isolate contains 37 repeats plus 4 other identical repeats composed of ASN-VAL-ASP-PRO (NVDP) at units 2, 4, 6 and 22. Therefore, at the beginning repeat containing-region of the protein, there exists an alternation between the basic repeat NANP and NVDP. The biological significance of the presence of these repeat analogues is unknown.

Naturally occurring polyclonal antibodies and monoclonal antibodies directed against the repeat region can inhibit the attachment of the sporozoites to liver cells (3,4). Similarly, antibodies directed against the synthetic peptide containing 3 repeats or a repetitive recombinant peptide were also able to inhibit the infectivity of the sporozoites (5,6). Hence, many research groups have concentrated on the synthetic peptides

Figure 1. Amino acid sequence of P. falciparum circumsporozoite protein.

MMRKLAILSVSSFLFVEALFNEYNCYGSSSNTRVLNELNYDNAGTNLYNE 50
LEMNYYGLNENWYSLKKNSRSLGENDDGNNNNGDNGREGKDEDKRDGNNE 100
DNEKLRKPKHKKLKQPGDGNPDPNANPNVDPNANPNVDPNANPNVDP 147
$(\overline{\underline{\text{NANP}}})_{19}$ NVDP $(\overline{\underline{\text{NANP}}})_{15}$ NKN 280
NQGNGQGHNMPNDPNRNVDENANANNAVKNNNNEEPSDKHIEQYLKKIKN 330
SISTEWSPCSVTCGNGIQVRIKPGSANKPKDELDYENDIEKKICKMEKCS 380
SVFNVVMSSIGLIMVLSFLFLN 402

The amino acid sequence was obtained from the corresponding DNA sequence (1,2).

of the repeat region as potential vaccines against P. falciparum malaria.

We have focused our attention on the immunogenicity of synthetic peptides of different length in mice. The peptides that were synthesized are described in Figure 1. It was found that $(NANP)_{40}$ could elicit an antibody response only in C57BL/6 mice, out of 9 strains of mice tested (7). In agreement with this observation, a T cell response to $(NANP)_{40}$ was obtained also in C57BL/6 mice (8). In contrast, C3H and DBA/2 mice did not respond.

When synthetic peptides of different lengths (Figure 2), i.e. $(NANP)_{20}$, $(NANP)_4$ and $Tyr(NANP)_3$, were used as immunogen in C57BL/6 mice, all of them gave rise to a T -cell response, and could be used as well as antigen in vitro, in any desired combination. On the contrary, an antibody response was only obtained when $(NANP)_{40}$ or $(NANP)_{20}$ was used as immunogen (8).

When the fine specificity of the T-cell response was studied at the clonal level, two families of T cells were obtained: one which was positive for the anti-T cell receptor monoclonal antibody F23.1 (9), and one which was negative for the same Mab. This Mab is thought to be specific for the gene product of the V_{β} 8 gene family of the ß chain of the T-cell receptor

Figure 2. CS synthetic antigens used in vitro and in vivo assays.

$(NANP)_{40}$	Obtained by polymerization of the activated ester tetrapeptide (NANP) Isolation by gel chromatography
$(NANP)_{20}$	As above
$(NANP)_{4}$	As above
$Tyr-(NANP)_{3}$	− NANP + Stepwise synthesis
$(NANP)_3$-NA	Stepwise synthesis, indicated as $(NANP)_3$
$(NANP)_2$-NA	Stepwise synthesis, indicated as $(NANP)_2$

and reacts with about 20% of spleen T cells. In our case, 5 out of 15 clones tested were positive for the F23.1 marker.

A further distinction can be made on the basis of the reactivity of the different clones toward different synthetic antigens, i.e. $(NANP)_4$, $(NANP)_3$, $(NANP)_2$ and Tyr-$(NANP)_3$.

If the concentration at which 50% of a proliferative response is obtained is considered as a criterium of specificity and relative strength of antigenic recognition among different clones, all the tested clones are different with respect to the amount of antigen needed to obtain 50% of the maximum proliferation and/or the relative potency among the four antigens used (Table 1). In particular, line 10 is not activated by $(NANP)_2$, while several other clones are (8). Or, in the case of clone 5/25, $(NANP)_3$ is a better antigen than Tyr-$(NANP)_3$, while the contrary is true in most of the other clones tested.

It is interesting to note that the addition of Tyr to $(NANP)_3$ is accompanied in general by a reduction in the dose of antigen required to obtain 50% of maximal proliferation. The reasons for these results are not

Table 1. Antigen concentration needed to obtain 50% proliferation of CS specific T cell clones or lines.

Clone	Antigen concentration (μg/ml) (NANP)$_4$	(NANP)$_3$-Tyr	(NANP)$_3$
11/3B	12.5	3.1	50
5/5H	1.6-3.1	0.8	1.6
5/9H	6.2	0.8	1.6
line 10	6.2	1.6	100
B2 +	6.2	6.2	25
line 5	3.2	6.2	1.6

The antigen concentration at which 50% of the maximum proliferation was obtained is shown. Proliferation of different clones ranged from 25 to 250 x 10^3 cpm. The concentration of the most active antigen is underlined.

clear at the moment, but could be ascribed to a better interaction of the peptide with the membrane due to the hydrophobic character of the tyrosine side chain or to the stabilization of the overall structure of the peptide.

The fact that several T-cell lines or clones responded poorly to (NANP)$_2$ while they did respond to (NANP)$_3$ gave us the opportunity to study the role of individual amino acids in the restoration of the response by testing the synthetic intermediates from (NANP)$_2$ to (NANP)$_3$.

In the case of line 10, it was found that the addition of a Proline residue to (NANP)$_2$ was sufficient to restore the response. Addition of the other 3 amino acid residues altered only slightly the dose response curve.

We believe that the effect of the addition of a Proline residue is a structural one such as to confer some rigidity to the rest of the peptide and not that this amino acid is of particular importance in the recognition process by the T-cell. Similar observations were obtained following analysis of the fine specificity patterns of additional clones. Therefore, the possibility exists that the peptide $(NANP)_n$ is exposed on the cell surface of the antigen presenting cells in a variety of different conformational structures, which in turn are recognized by T cells. Alternatively, the peptide exists in a preferred conformation but it may undergo multiple conformational changes upon interaction with the T cell receptor and class II molecules.

CONCLUSION

The strict genetic restriction observed for the immune response of the synthetic peptide $(NANP)_{40}$ in mice raises the question of whether or not a similar genetic restriction may be observed in humans. If this turns out to be true, the onset of natural protection acquired by the individual in endemic areas might be linked to a higher degree of responsiveness to the CS protein. On the other hand, immunization of mice with the $(NANP)_{40}$ synthetic peptide gave rise to a remarkably heterogenous T cell response, in spite of its simple amino acid composition.

REFERENCES

1. Dame, J.B., Williams, J.L., McCutchan, T.F., Weber, J.L., Wirtz, R.A., Hockmeyer, W.T., Maloy, W.L., Haynes, J.D., Schneider, I., Roberts, D., Sanders, G.S., Reddy, E.P., Diggs, C.L. and Miller, L.H. Science, 225, 593-599.
2. Enea, V., Ellis, J., Zavala, F., Arnot, D.E., Asanavich, A., Masuda, A., Quakyi, I. and Nussenzweig, R.S. Science 225, 628-630.
3. Hollindale, M.R., Nardin, E.H., Tharvanij, S., Schwartz, A.L. and Nussenzweig, R.S. J. Immunol. 132, 909-913.
4. Nussensweig, V. and Nussenzweig, R.S. in: Synthetic peptides as antigens, pp. 150-163 (Ciba Foundation Symposium 119) Wyley, Chichester.
5. Ballou, W.R., Rothbard, J., Wirtz, R.A., Gordon, D.M., Williams, J.L., Gore, R.W., Schneider, I., Hollingdale, M.R., Beaudoin, R.L., Maloy, W.L., Miller, L.H. and Hockmeyer, W.T. Science 228, 996-999.

6. Mazier, D., Mellouk, S., Beaudoin, R.L., Texier, B., Druilhe, P., Hockmeyer, W., Trosper, J., Paul, C., Charoenvit, Y., Young, J., Miltgen, F., Chedid, L., Chigot, J.P., Galley, B., Brandicourt, O. and Gentilini, M. Science 231, 156-159.
7. Del Giudice, G., Cooper, J.A., Merino, J., Verdini, A.S., Pessi, A., Togna, A.R., Engers, H.D., Corradin, G. and Lambert, P.H. J. Immunol., in press.
8. Togna, A.R., Del Giudice, G., Verdini, A.S., Bonelli, F., Pessi, A., Engers, H.D. and Corradin, G. J. Immuno., in press.
9. Staerz, V.D., Rammensee, H.G., Benedetto, J.D. and Bevan, M.J. J. Immmunol. 134, 3994-4000.

MAPPING OF ANTIGEN EPITOPES INTERACTING WITH CLASS II MHC PRODUCTS AND WITH THE ANTIGEN RECEPTOR OF T LYMPHOCYTES

L. Adorini, C. Bove, M. Darsley*, E. Appella*, G. Doria

Laboratory of Pathology, ENEA C.R.E. Casaccia, C.P. 2400, 00100 Rome A.D., Italy; * Laboratory of Cell Biology, National Cancer Institute, NIH, Bethesda, MD 20892, U.S.A.

INTRODUCTION

Helper T cells recognize antigen in association with class II MHC products (Ia molecules) on the surface of antigen-presenting cells (1, 2). Structural studies have demonstrated that the antigen receptor of both T and B cells is encoded by variable, diversity, joining and constant gene segments rearranged during ontogeny to generate approximately the same diversity of ligand binding sites (3). Unlike immunoglobulins, recognizing soluble, unprocessed protein antigens, the antigen receptor of T lymphocytes recognizes a bimolecular complex composed of class II MHC molecules and an antigen fragment derived from proteolytic digestion of the native molecule (4, 5). The simultaneous recognition by a single receptor of antigen associated to Ia molecules indicates that for effective T cell recognition antigen and Ia molecules must physically interact on the membrane of antigen-presenting cells (6). This interaction can explain the Ir gene control of immune responsiveness, whereby certain antigen epitopes are presented effectively only by particular Ia molecules (7), and the determinant selection hypothesis postulating a selection by antigen-presenting cells of the epitope to be presented to T cells (8). Recent reports directly support the antigen-Ia interaction model but it is still debated whether this interaction is stable and occurring in the absence of T cells (9) or whether it is transient and stabilized by the T cell receptor (10). A corollary of this model is that distinct antigen epitopes interact with Ia molecules and with the T cell receptor, as indicated by studies on the recognition of cytochrome c by T cells (11, 12).

In the present study we have examined the antigen specificity of a T cell hybridoma obtained by fusion of hen egg-white lysozyme (HEL)-primed T lymphocytes from (C57BL/10xDBA/2)F1 mice with T cell lymphoma cells BW5147. This hybridoma, 3A1, is activated by HEL and by the synthetic peptide 105-120 of HEL in association with I-E^d molecules to produce interleukin 2 (IL-2). Amino acid residues of the peptide 105-120 involved in activation of 3A1 cells have been identified by the reactivity of 3A1 hybridoma to lysozymes and to synthetic analogues of truncated peptides of the HEL region 105-120.

Characterization of the 3A1 hybridoma. The 3A1 T cell hybridoma was obtained by fusion of lymph node cells from (C57BL/10xDBA/2)F1 (BDF1) mice, with the T cell lymphoma line BW5147. Popliteal lymph node cells were obtained from mice immunized 8 days before into the hind footpads with 50 μg HEL-CFA and restimulated *in vitro* for 3 days with 100 μg/ml HEL. After fusion cells were cultured in selective medium and growing hybrids were screened for reactivity to HEL by HEL-dependent IL-2 production. Microcultures containing $2x10^4$ hybridoma cells and $5x10^5$ syngenic irradiated (3000 R) spleen cells as a source of antigen-presenting cells were set up in 0.2 ml of culture medium. After 24 hours of culture 0.1 ml of supernatant medium was transferred to microculture wells containing $1x10^4$ CTLL cells in 0.1 ml medium. After additional 24 hour incubation the presence of IL-2 was detected by tritiated thymidine incorporation by CTLL cells. Hybridomas producing IL-2 in response to HEL were then screened against a panel of HEL synthetic peptides encompassing almost all the lysozyme molecule. Hybridomas reactive to HEL and to HEL peptides were cloned by limiting dilution. After the initial screening, IL-2 production was quantitated by tritiation of samples and probit transformation of data as previously described (13). The 3A1 hybridoma, reactive to HEL and to the HEL synthetic peptide 105-120, has been analyzed in detail. This hybridoma recognizes antigen in the context of I-E^d molecules as demonstrated by the ability of BDF1 and DBA/2 antigen presenting cells (APC) to induce IL-2 production, whereas APC from B6 or A/J mice fail to do so, and by inhibition of IL-2 production by addition to cultures of anti-I-E^d monoclonal antibodies. The 3A1 hybridoma displays, on a molar basis, a higher reactivity to the HEL peptide 105-120 than to the native HEL molecule. Maximal IL-2 production is induced by 20 μM HEL or by 2 μM peptide 105-120 and half-maximal IL-2 production is induced by about 7 μM HEL or 0.7 μM peptide 105-120. Higher reactivity to a given peptide than to the native molecule has been observed for many T cell hybridomas and clones indicating that pre-processed antigen is presented more effectively by APC to T cells.

The T cell receptor for antigen expressed by 3A1 cells is recognized by the monoclonal antibody KJ-16.133, specific for a conserved epitope of the Vß8 gene product, expressed by about 20% of T lymphocytes (14). This has been demonstrated by inhibition of 3A1 cell activation after addition to cultures of monoclonal antibody KJ-16.133. The 3A1 hybridoma also expresses the L3T4 antigen, a marker of helper T lymphocytes, as revealed by cyto-fluorimetric analysis with the monoclonal antibody GK1.5. However, addition to cultures of GK1.5 antibody does not induce a complete inhibition of 3A1 cell activation by HEL and DBA/2 APC.

Activation of 3A1 cells by native lysozymes and by HEL synthetic peptides. As already mentioned, 3A1 hybridoma cells, tested against a panel of 10 HEL synthetic peptides, are activated only by peptide 105-120. X-ray crystal-lography of HEL (15) shows that peptide 105-120 has, as compared to the entire HEL sequence, a higher than average frequency of external amino acids, a -helical structure from residue 108 to 115, and an average thermal mobility. Hydrophilicity analysis based on the Hopp and Woods scale (16) indicates that peptide 105-120 contains the highest hydrophilicity peak along the HEL sequence. The HEL region 105-120 does not possess an amphipathic structure with -helical periodicity, a charac-teristic correlated to antigenic sites recognized by T cells (17), but it displays the highest peak of differen-tial amphipathicity. This region of the HEL molecule has already been shown to represent an immunodominant epitope for B10.D2 (H-2^d) T lymphocytes (18).

	105 110 115 120	3A1 activation
Mouse	IR------AH-QNRL-	
HEL	MNAWVAWRNRCKGTDV	+
T-HEL	---------------- (↓ ↓ ↓)	-
REL	--------KH------	-
DEL II	-----------R----	+
HUL	IR---------QNR--	-

FIGURE 1. Recognition by 3A1 hybridoma of HEL, tryptic digest of HEL and variant lysozymes. 3A1 activation refers to the production of IL-2 by 3A1 cells stimulated with irradiated DBA/2 spleen cells and HEL, tryptic digest of HEL (T-HEL), ring-necked pheasant lysozyme (REL), duck lysozyme (DEL II) or human lysozyme (HUL). The HEL sequence 105-120 is indicated by the single letter amino acid code. In variant lysozymes as well as in mouse lysozyme residues shared with HEL are indicated by a dash.

To identify the residues of peptide 105-120 involved in 3A1 cell activation we have examined first the ability of this T cell hybridoma to recognize denatured HEL or variant lysozymes. As indicated in Fig. 1, 3A1 activation is induced by HEL and duck lysozyme (DEL II) whereas a total tryptic digest of HEL (T-HEL), ring-necked pheasant lysozyme (REL) or human lysozyme (HUL) fail to activate 3A1 cells to produce IL-2.

These results indicate that residues 113 and 114 are critical for 3A1 cell activation as demonstrated by the failure of REL to induce IL-2 production by 3A1 cells. In addition, other residues located in the N-terminal and/or the C-terminal portion of the region 105-120 play a role in antigen recognition by 3A1 cells, as indicated by the failure of HUL to induce IL-2 production.

On the other hand, a K → R substitution at residue 116, as in DEL II, does not substantially affect the ability of 3A1 cells to recognize antigen. However, use of variant native molecules to map antigen epitopes involved in T cell activation may be misleading since it has been demonstrated that amino acid residues distant from the antigenic epitope may affect antigen handling by APC (19). This problem is completely eliminated by using synthetic peptides. Synthetic peptides of the HEL sequence 105-120 substituted at residues 113, 105-120 (K113), or 114, 105-120 (H114) do not induce IL-2 production by 3A1 hybridoma indicating that these two residues are both critical for 3A1 cell activation. Conversely, a synthetic peptide substituted at position 118, 105-120 (R118), is almost as effective as the 105-120 peptide in inducing IL-2 production indicating that this non-conservative T → R substitution does not affect 3A1 cell activation.

Mapping of amino acid residues interacting with $I\text{-}E^d$ molecules and with the 3A1 T cell receptor. Antigenic peptides do not stimulate T cells in the absence of appropriate MHC class II-bearing APC indicating that both antigen and Ia molecules are involved in T cell activation. Results by Babbitt et al. (9) measuring the binding of a HEL antigenic peptide to purified I-A molecules have demonstrated that the peptide antigen has a measurable affinity for I-A molecules from responder mouse strains but does not bind to the I-A molecule of a non-responder strain. These results agree with observations demonstrating that related synthetic amino acid polymers (20, 21) or synthetic peptides (22) compete for presentation by APC to T cells. We have examined the ability of peptides 105-120 (K113) and 105-120 (H114), both non-stimulatory for 3A1 cells, to inhibit the presentation of the unsubstituted peptide 105-120. Results in Fig. 2 demonstrate that peptide 105-120 (K113) competes very efficiently with peptide 105-120 in binding to $I\text{-}E^d$ molecules on the surface of APC. This observation indicates that these two peptides share the epitope interacting with $I\text{-}E^d$ molecules and thus the lack of activation of 3A1 cells by peptide 105-120 (K113) defines an amino acid residue involved in interaction with the T cell receptor. Conversely, the peptide 105-120 (H114) not only fails to activate 3A1 cells, but also it does not compete with peptide 105-120 for 3A1 cell activation

Peptide	94 100 105 110 115 120 CAKKIVSDGNGMNAWVAWRNRCKGTDV	3A1 activation	competition for 105-120
105-120 (K113)	--------K-------	-	+
105-120 (H114)	---------H------	-	-
105-120 (R118)	-------------R--	+	
94-110	--------------	-	-

FIGURE 2. Activation of 3A1 cells by analogues of peptide 105–120 and their competition with peptide 105–120 for binding to APC. Peptide 105–120 (R118) which activates 3A1 cells has not been tested in the competition assay.

indicating that the amino acid residue at position 114 is implicated in binding of the peptide to the I-E^{d} molecules. The finding that two adjacent amino acid residues interact one with the T cell receptor and the other with Ia molecules may indicate that these two residues are part of distinct, but contiguous antigen epitopes. Alternatively, it is possible that amino acid residues contacting the T cell receptor and Ia molecules do not constitute discrete epitopes but that antigenic residues interacting with these structures are interspersed in the antigen molecule.

It is interesting to note that the N → K substitution at position 113 is semi-conservative according to the tabulation by Grantham (23) and that both residues are polar. The R → H substitution at position 114 is conservative but R is polar and charged whereas H is neutral. Thus, it is likely that the interaction between peptide antigen and Ia molecules depends primarily on hydrophobic non-covalent interactions whereas in the interaction between antigen and the T cell receptor a closer contact is required among the lateral chains of the amino acid residues involved. Babbitt et al. (24) have recently demonstrated that synthetic peptides non-stimulating for T cell hybridomas compete for binding and inhibit presentation of immunogenic peptides. No evidence was found in these studies for laterally separated functional areas contacting Ia and the T cell receptor but rather an intermingling, on the antigenic peptide, of residues contacting these two molecules, as already suggested by analysis of the T cell response to cytochrome peptides (25). Further biochemical and functional studies will clarify this issue but a complete understanding of the interaction among antigen, Ia molecules and the T cell receptor will probably be achieved only by X-ray crystallographic analysis of this trimolecular complex.

REFERENCES

1. Rosenthal, A.S. and Shevach, E.M. (1973) J. Exp. Med. 138, 1194.

2. Schwartz, R.H. (1985) Ann. Rev. Immunol. 3, 237.
3. Kronenberg, M., Siu, G., Hood, L. and Shastri, N. (1986) Ann. Rev. Immunol. (in press).
4. Unanue, E.R. (1984) Ann. Rev. Immunol. 2, 395.
5. Shimonkevitz, R., Colon, S. Kappler, J.M.,Marrack, P. and Grey, H.M. (1984) J. Immunol. 133, 2067.
6. Kappler, J.W., Skidmore, B., White, J. and Marrack, P. (1981) J. Exp. Med. 153, 1198.
7. McDevitt, H.O. and Benacerraf, B. (1969) Adv. Immunol. 11, 31.
8. Rosenthal, A.S., Barcinski, M.A. and Blake, J.T. (1977) Nature (London) 267, 150.
9. Babbitt, B.P., Allen, P.M., Matsueda, G., Haber, E. and Unanue, E.R. (1985) Nature 317, 359.
10. Watts, T.H., Gaub, H.E. and McConnell, H.M. (1986) Nature 320, 179.
11. Hansburg, D., Fairwell, T., Schwartz, R.H. and Appella, E. (1983) J. Immunol. 131, 319.
12. Heber-Katz, E., Hansburg, D. and Schwartz, R.H. (1983) J. Mol. Cell. Immunol. 1, 3.
13. Sette, A., Adorini, L., Marubini, E. and Doria, G. (1986) J. Immunol. Methods 86, 265.
14. Haskins, K., Kubo, R., White, J., Pigeon, M., Kappler, J. and Marrack, P. (1983) J. Exp. Med. 157, 1149.
15. Artymiuk, P.J., Blake, C.C.F., Grace, D.E.P., Oatley, S.J., Phillips, D.C. and Sternberg, M.J.E. (1979) Nature 280, 563.
16. Hoop, T.P. and Woods, K.R. (1981) Proc. Natl. Acad. Sci. USA 78, 3824.
17. De Lisi, C. and Berzofsky, J. (1985) Proc. Natl. Acad. Sci. USA 82, 7048.
18. Katz, M.E., Maizels, R.M., Wicker, L., Miller, A. and Sercarz, E.E. (1982) Eur. J. Immunol. 12, 535.
19. Shastri, N., Miller, A. and Sercarz, E.E. (1986) J. Immunol. 136, 371.
20. Werdelin, O. (1982) J. Immunol. 129, 1833.
21. Rock, K.L. and Benacerraf, B. (1983) J. Exp. Med. 157, 1618.
22. Lakey, E.K., Margoliash, E., Flouret, G. and Pierce, S.K. (1986) Eur. J. Immunol. 16, 721.
23. Grantham, R. (1974) Science 184, 862.
24. Babbit, B.P., Matsueda, G., Haber, E., Unanue, E.R. and Allen, P. (1986) Proc. Natl. Acad. Sci. USA 83, 4509.

25. Hansburg, D. and Appella, E. (1985) J. Immunol. 135, 3712.

SECTION IV

ANALYTICAL AND PREPARATIVE BIOAFFINITY METHODS

ANALYTICAL USE OF ANTIBODIES BASED ON SPECIFIC MOLECULAR BINDING

A. Albertini, C. Iacobello, G. Ruggeri and P. Arosio

Institute of Chemistry, Univ. of Brescia, 25100 Brescia, Italy.
Dept. of Biomedical Science and Technology, Univ. of Milan, 20132 Milan, Italy.

In the last few years the performance of analytical systems based on antibodies has improved enormously, due to the simplification of techniques, the development of non-isotopic tracers to follow the antigen-antibody reaction and the introduction of monoclonal antibodies with better defined specificity. An increasing number of fields in medicine are now interested in the development of immunoassays.

Ideally, an immunoassay should be highly specific for the analyte(s) to be measured, have a sensitivity sufficient to detect the analyte(s) in all the specimens and a broad enough measurement range to cover all the concentrations occurring in the samples, express the concentration of the analyte(s) in biologically significant units, and be technically fit, precise, fast, simple and economic. To fulfill all these requirements it is necessary to: a) know in detail the biochemical and biological characteristics of the analyte(s); b) obtain ligands (antibodies) with high affinity and specificity for the analyte(s); c) assess the effects of the various types of immunoassay on the immunological reaction and its response.

The immunological assays are the systems of choice for the evaluation of proteins, because of the facility to obtain specific antibodies that can be used for the development of highly sensitive radio- or enzyme-immunoassays.

In order to improve the performance and significance of the immunoassays for protein evaluation, the antigen-antibody reaction needs to be studied in detail, particularly for proteins with a complex quaternary structure. Following this line of thought, we have been using isoferritins as model proteins since they offer a number of interesting characteristics: the three-dimensional structure is known in detail so that it is possible to analyze the data at a molecular level; they are highly immunogenic; and, moreover, the evaluation of ferritin in the serum is a significant index of the body's iron stores, and is useful for the diagnosis of anemias and other pathologies (1).

ISOFERRITIN STRUCTURE

Isoferritins are iron storage proteins found in all tissues. They are composed of 24 subunits which assemble in an almost spherical hollow shell of about 500,000 Mr (2); in turn, the subunits are made up of four tightly packed alpha helices (2). Ferritins have a high stability towards heat and chemical agents, and can be dissociated only at extreme pH values and temperatures (2). Human ferritins contain two subunit types, called L and H, of 174 and 182 amino acids, respectively (3,4). The cDNA and genes coding for the L and H subunits have been cloned and sequenced (5,7); the two peptides have about 55% sequence homology, with 25% non-conservative substitutions (8,9). Much of the complexity of these proteins arises from the fact that both subunits are always present in different proportions in all natural ferritins, thus giving rise to a large degree of heterogeneity (3).

Minor, but significant amounts of ferritins are also found in serum and in all body fluids tested so far (10); they probably arise from active secretion from the various cell types. Serum ferritin is glycosylated in high proportion (70-80%) as shown by its binding to the lectin Concanavalin A (11). This ferritin contains an additional subunit of 23,000 Mr, which is glycosylated, and is immunochemically similar, but not identical, to the L subunit in the tissues (12).

IMMUNOLOGICAL METHODS FOR FERRITIN EVALUATION

Most of the immunological methods developed for ferritin evaluation are aimed at the measurement of serum ferritin which is present at concentrations ranging from 15 to 250 ug/l normally, and up to 20,000 µg/l in some pathologies (13). The majority of the available assays are based on ferritins extracted from liver or spleen, which have immunochemical properties similar, but not identical, to serum ferritin (14).

The assays so far developed use an isotopic or enzymatic label, and are based either on the competition of the unknown with the label antigen for binding to a limited amount of antibody (RIA systems), or on the binding of the antigen to solid phase antibody, followed by the reaction with the labeled antibody (sandwich systems). The commercial ferritin assays use these two different systems with the assumption that they give similar results. In order to test the validity of this assumption, we developed various ferritin assays, using the same antibody in conjunction with RIA and sandwich systems. The responses of the assays were highly correlated, indicating that they measure the same molecules. However, the RIA methods over-estimated serum ferritin by 30-40% in comparison to the sandwich methods (Table 1). This difference can easily be corrected by the use of proper standards. In any case, this finding indicates that the immunoassay system may have a significant effect on the evaluation of the analyte, and suggests that the molecule to be analysed present in the serum is not immunochemically identical to the one of the standards (extracted from liver). The difference appears to be enhanced in systems where these two ferritins compete for binding the antibody (15).

One of the problems in most immunological assays is the large variability of analyte concentration detected by different assays, which renders the comparison of data from various laboratories difficult. The different characteristics of the reagents (antibodies and antigens) and of the assay architecture can be expected to play a major role in this variability, but the most important factors are not known.

Table 1. Evaluation of serum ferritin with five different methods based on the same antibody and standards.

System	Response	Correlation (r)
IRMA	100%	1.00
ELISA.POD	82%	0.99
ELISA.GAL	83%	0.99
RIA.Lis	152%	0.99
RIA.D.A.	132%	0.99

IRMA: sandwich using radiolabeled antibody; ELISA.POD: sandwich using horse radish peroxidase as tracer; ELISA. GAL: sandwich with beta-galactosidase; RIA.Lis: RIA with a solid phase second antibody for the separation of bound antigen; RIA.D.A: RIA with a second antibody for the separation of bound antigen.

To this end we studied a number of commercial serum ferritin immunoassays in terms of their ability to evaluate three different ferritin types: a liver ferritin reference preparation, a calibrated heart ferritin sample and serum ferritin. In all cases the responses were highly correlated, but the concentrations detected varied in the different kits. No simple cause for this discrepancy was evident. After standardization of the kits with the same reference preparation, a marked decrease of the variability for serum ferritin determination was observed, particularly in the lower range of concentrations. Moreover, it appeared that the degree of cross reactivity of the antibodies with tissue ferritins rich in the H subunit (i.e., from the heart) did not influence the evaluation of serum ferritin, in keeping with the finding that in serum ferritin the concentration of the H subunit is negligeable (14).

Immunological Heterogeneity of Isoferritins. The minor immunochemical differences between serum and liver ferritins do not significantly affect the clinical importance of serum ferritin determination. On the contrary, the isoferritins extracted from different tissues have strikingly different immunochemical charac-

Table II. Ferritin concentration detected by various commercial kits.

Kit	Source of ferritin		
	Liver (a)	Heart (a)	Serum (b)
A	100%	12%	95%
B	124%	14%	98%
C	108%	126%	91%
D	101%	15%	106%
E	193%	20%	178%
F	90%	18%	114%
G	71%	23%	96%

(a) Calculated over protein concentration
(b) Calculated over consensus mean

teristics, that have to be taken into account to evaluate their concentration. The first evidence of the immunochemical heterogeneity of isoferritins was obtained with antisera elicited by ferritin from liver (rich in the L subunit) and from HeLa cells (rich in the H subunit) (16), which were specific for their own immunogen while reacting weakly with the other ferritins. These antisera were used to develop immunoassays that allowed the distinction and quantitation of the "acidic" (rich in H subunit), and the "basic" (rich in L subunit) isoferritins (17).

Recently, we developed a library of monoclonal antibodies for human isoferritins in order to obtain better reagents for the quantitation of acidic and basic isoferritins, and to evaluate the possibility of producing antibodies that would quantitate all human isoferritins irrespective of their subunit composition. Ferritins from liver (90% L subunit) and heart (70% H subunit) were employed as immunogens and for screening the antibodies. Four independent systems were used to detect the useful hybridoma cell lines; all gave comparable results. From this preliminary analysis it appeared that absorption to polystyrene plates did not affect the immunochemical properties of ferritin, and that most of the antibodies were able to produce precipitin lines in double diffusion (Ouchterlony) experiments, in keeping with the highly repetitive structure of the antigens (18, 19).

The antibodies showed high binding only to one of the two immunogens, apart from two which appeared to recognize both immunogens. However, further studies showed that these two antibodies, when used in a sandwich immunoassay, also give stronger signals with one of the two antigens. These findings indicate that it is unlikely that there are epitopes common to the H and L subunits. Probably, the quantitation of all isoferritins can be approached only through the use of a proper cocktail of antibodies with different specificities.

Table III. Screening of anti-ferritins monoclonal antibodies.

Immunogen (a)	n.	ELISA (b)		Immuno-precipitation(c)		Ouchterlony (d)
		LF	HF	LF	HF	
LF	8	++	-	75-100	-	LF
HF	8	-	++	-	37-100	HF
HF	6	++	-	36-67	-	LF
HF	3	-	+	-	53-83	-
HF	2	+	+	46-61	29-78	-

(a) LF: liver ferritin; HF: heart ferritin.
(b) Response in ELISA system in which the antibodies were reacted with liver (LF) or heart (HF) ferritin absorbed on polystyrene plates. + for positive, - for negative.
(c) Percentage of precipitation of radiolabeled liver (LF) or heart (HF) ferritins with antibody in excess.
(d) Antigen producing precipitin lines.

Liver ferritin elicited antibodies of similar specificity, while among the antibodies elicited by the heart preparation 42% had a specificity similar to the anti-liver antibodies. This unexpected result shows that the two antigens have different distributions of the same epitopes, and that the ones predominant in liver are more immunogenic.

Localization of Antibodies Binding Sites. In order to map the antibody epitopes on the ferritin molecule, we performed immunoblotting experiments on the ferritin subunits separated by SDS polyacrylamide gel electrophoresis. All the antibodies were able to bind at least one subunit, but their affinity for denatured ferritin decreased dramatically: a clear signal was obtained only if the denaturing gels were loaded at least 100 times more than the non-denaturing gels. The results from the blotting experiments show that all the antibodies that bind preferentially liver ferritin, recognize the L chain, while the antibodies that bind heart ferritin also bind the H chain. Three antibodies were able to recognize both subunits. The large decrease of immunoreactivity of denatured ferritin indicates that only a fraction of the epitopes is sequential, and that the three-dimensional structure is essential for the high affinity binding, in agreement with results obtained on other proteins (19). This finding makes the detailed mapping of the epitopes difficult, because of the poor reaction of the antibodies with ferritin peptides.

The location of the epitopes was also approached by competition experiments with antibodies with known specificity for the subunits. All the antibodies could be divided into two classes: one competing with the 2A4 antibody (specific for the H chain), and the other competing with LO3 anti-liver antibody (specific for the L chain) (18). The antibodies belonging to the same class competed with each other, but not with the antibodies from the other class. Therefore, the isoferritin molecules appear to have two types of epitopes grouped in close proximity, respectively located on the H and L subunit.

On the basis of the ferritin structure one can postulate that the epitopes are located on the subunits, and probably not at the interface between them. The

Table IV. Localization of epitopes.

Immunogen	n.	Subunit bound in W. Blotting (a)	Competition (b) aLF	aHF
LF	8	L	+	–
HF	8	H	–	+
HF	6	L(H)	+	–
HF	3	H	–	+
HF	2	H,L	+	+

(a) L for L subunit; H for H subunit.
(b) results in competition experiments with antibodies specific for the L subunit (aLF) and for the H subunit (aHF).

area of the ferritin subunit exposed on the outer surface of the molecule is about 25 x 55 Å, which is slightly larger than the area occupied by the binding of an antibody (19). Therefore, it is conceivable that each subunit may not bind more than one antibody at the same time. Moreover, the H and L subunits, despite their common

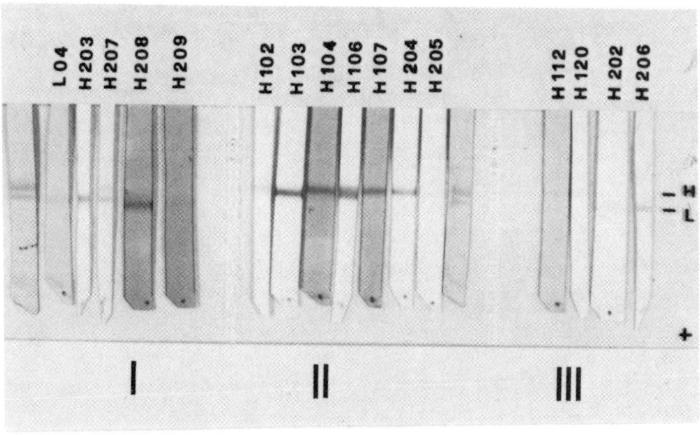

Figure 1. Immunoblotting of the H and L subunits from heart ferritin with various antibodies.

overall structure (21), have large differences in the surface amino acids where no common sequences of more than 4 amino acids can be observed (22). It is possible that antibodies which are highly specific for one of the two chains may bind to both chains under conditions in which the structure of the epitopes is altered and the affinity depressed.

Effect of the Methods on Antibody Specificity. The logical use of monoclonal antibodies is the development of specific immunoassays that could be used to study the biology of the analyte. In addition, they are probes for the two subunits, and as such represent interesting tools to study the disposition of the two subunits in the quaternary structure of the molecule, a task that cannot be accomplished by crystallography (21).

We developed an IRMA assay based on an anti-H subunit antibody (2A4). To our surprise, this assay was not able to detect liver ferritin despite the presence of

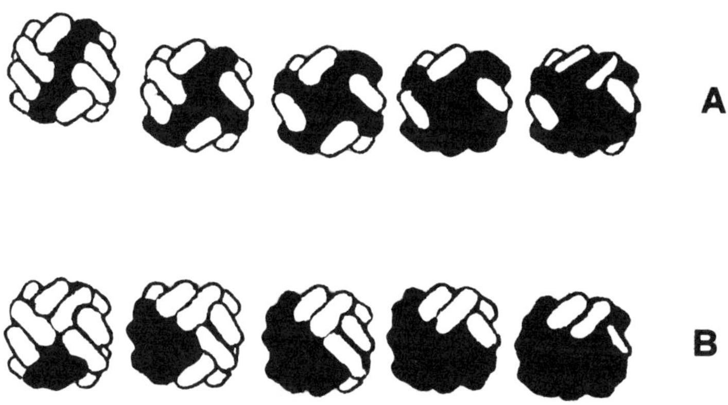

Figure 2. Models for the quaternary structure of isoferritins. The two subunits may be randomly distributed in the molecules as in (A) or form clusters as in (B). Our data indicate that B is a more accurate representation of the isoferritins.

Table V. Ferritins detected in various assay types by monoclonal antibodies.

Assay	Amount detected	
	Liver Ft by anti H-chain	Heart Ft by anti L-chain
Sandwich	<0.01%	8%
RIA	<0.01%	3%
Dotting	5%	50%

10% of the H chain (20). In one of the methods used for screening the hybridomas, this antibody was able to recognize liver ferritin bound to the solid phase. Similar results were obtained with other antibodies for the H subunit. In sandwich assays, the anti-L chain antibodies detected only 8% of the heart ferritin, despite the presence of about 30% L chain. This shows that the system used in the immunoassay has a strong effect on the apparent specificity of the antibodies (18).

Using a quantitative dotting system, in which the antigen is absorbed on the solid phase, we could obtain a good agreement between the observed signal and the subunit composition of the samples.

We then separated isoferritins with different subunit composition, and we could see that the anti-H chain sandwich assay was able to detect only the isoferritins with H subunit content above 50% (23).

These findings give indirect information on the structure and immunogenicity of isoferritins. They suggest that the higher the density of epitopes on a molecule the higher the affinity for the antibody, as shown by the RIA experiments. Moreover, the striking difference between the binding of a labeled antibody to ferritin caught by the ligand, or absorbed on a solid phase suggest that the solid phase antibody orients the molecules so that the epitopes are not available to the second antibody unless the content of the epitope is sufficiently high (>50%). One possible and likely explanation of this finding is that the two subunits form clusters in the molecule (Figure 2).

CONCLUSIONS

In this work isoferritins were used as model molecules to study the effects of various immunoassay types on the quantitation of an analyte and on the specificity of antibodies. The results indicate that the architecture may influence the results of the assay, and that probably the major cause of variability among different assays is an improper standardization. The use of different assay types may modulate the specificity of the antibody; for instance some molecules that are not detected by radioimmunological methods can be recognized in systems similar to those used in immunohistochemistry.

Acknowledgements. This work was partially supported by CNR Progetti Finalizzati Tecnologie Biomediche. We are grateful to Dr. E. Brocchi, L. Cappucci, A. Cozzi, F. de Simone, D. Gamba, S. Levi, and A. Luzzago.

REFERENCES

1. Lipschitz, D.A., Cook, J.D. and Finch, C.A. (1974) N. Eng. J. Med. 290, 1213.
2. Harrison, P.M., Clegg, G.A., May, K. (1980) in: Iron in Biochemistry and Medicine (Jacobs and Worwood, eds.) Academic Press, New York, p. 131.
3. Arosio, P., Adelman, T.G. and Drysdale, J.W. (1978) J. Biol. Chem. 252, 4451.
4. Stefanini, S., Chiancone, E., Arosio, P., Finazzi Agro', A. and Antonini, E. (1982) Biochemistry 21, 2293.
5. Costanzo, F., Santoro, C., Colantuoni, V., Bensi, G., Raugei, G., Romano, V. and Cortese, R. (1984) EMBO J. 3, 23.
6. Costanzo, F., Colombo, M., Staempfli, S., Santoro, C., Marone, M., Frank, R., Delius, H. and Cortese, R. (1986) Nucl. Acids Res. 14, 721.
7. Santoro, C., Marone, M., Ferrone, M., Costanzo, F., Colombo, M., Minganti, C. and Cortese, R. (1986) Nucl. Acids Res. 14, 2863.
8. Boyd, D., Vecoli, C., Belcher, D.M., Jain, S.K. and Drysdale, J.W. (1985) J. Biol. Chem. 260, 11755.

9. Dorner, M.H., Salfeld, J., Will, H., Leibold, E.A., Vass, J.K. and Munro, H.N. (1985) Proc. Natl. Acad. Sci. USA 82, 3139.
10. Ruggeri, G., Iacobello, C., Albertini, A., Brocchi, E., Levi, S., Gabri, E. and Arosio, P. (1984) in: Ferritins and Isoferritins as Biochemical Markers (Albertini et al., eds.) Elsevier, Amsterdam, pp. 67-79.
11. Carenini, A., Ronchi, D., Garbetta, F., Pozza, G., Gabri, E., Levi, S. and Arosio P. (1985) Clin. Chim. Acta 152, 165.
12. Santambrogio, P., Cozzi, A., Levi, S. and Arosio, P. (1987) Brit. J. Haematol. (in press).
13. Worwood, M. (1979) CRC Crit. Rev. Clin. Lab. Sci. 10, 171;
14. Iacobello, C., Belloli, S., Ghielmi, S., Albertini, A. and Arosio, P. (1984) Clin. Chem. 30, 298.
15. Ghielmi, S., Iacobello, C., Pizzoccolo, G., Albertini, A., and Arosio, P. (1982) Clin. Chim. Acta 120, 309.
16. Arosio, P., Yokota, M. and Drysdale, J.W. (1976) Cancer Res. 36, 1735.
17. Arosio, P., Iacobello, C., Montesoro, E. and Albertini, A. (1981) Immunol. Lett. 3, 309.
18. Luzzago, A., Arosio, P., Iacobello, C., Ruggeri, G., Capucci, L., Brocchi, E., De Simone, F., Gamba, D., Levi, S. and Albertini, A. (1986) Biochim. Biophys. Acta 872, 61.
19. Barlow, D.J., Edwards, M.S. and Thornton, J.M. (1986) Nature 322, 747.
20. Cavanna, F., Ruggeri, G., Iacobello, C., Chieregatti, G., Murador, E., Albertini, A. and Arosio, P. (1983) Clin. Chim. Acta 134, 347.
21. Bolognesi, M., Gatti, G., Levi, S., Gabri, E. and Arosio, P. (1984) FEBS Lett. 165, 63.
22. Harrison, P.M., White, J.L., Smith, J.M.A., Farrants, J.W., Ford, G.C., Rice, D.W., Addison, J.M. and Treffry, A. (1985) in: Proteins of Iron Storage and Transport (Spik et al., eds.) Elsevier, Amsterdam, p. 67.
23. Arosio, P., Levi, S., Gabri, E., Ruggeri, G., Iacobello, C., Albertini, A. and Cavanna, F. (1984) in: Ferritins and Isoferritins as Biochemical Markers (Albertini et al., eds.) Elsevier, Amsterdam, p. 33.

MODIFIED AVIDINS IN DETECTION SYSTEMS

F. Chiolerio*, E. Filippini*, S. Magnaghi*
G. Valentini° and M. Malcovati§

*Società Prodotti Antibiotici, Via Biella 8, Milan; °Dipartimento di Biochimica, Univ. di Pavia; §Dipartimento di Biologia e Genetica per le Scienze Mediche, Univ. di Milan.

The avidin-biotin system, because of avidin's ability to bind with high affinity up to four moles of biotin and because of the ease with which biotin can be bound to almost any class of biological compounds, has been indicated in recent years as one of the most powerful tools in bridging together different macromolecules. In particular, this system has been applied to the detection of biotinylated probes (antibodies, DNA) by enzyme-linked assays and to the immobilization of macromolecules onto solid phases.

The main drawback to this system lies in the fact that hen-egg avidin, owing to its very basic isoelectric point, binds aspecifically to negatively charged macromolecules (nucleic acids, mucopolysaccharides, etc.) in biological specimens.

In order to circumvent this difficulty, hen-egg avidin was reacted with different acidic anhydrides (acetic, succinic, phthalic) to obtain neutral or acidic derivatives. Such derivatives maintained their multimeric structure, as judged by gel filtration (Figure 1) and most of the original biotin-binding activity (Table 1).

The ability of such derivatives to bridge different macromolecules and in particular to form complexes

Table 1. Biotin-binding activity of native and modified avidins.

	U/mg protein	%
Native avidin	14.58	100
Acetyl avidin	10.6	72.7
Phthalyl avidin	6.4	43.9
Succinyl avidin	7.6	52.1

with a polybiotinylated enzyme (horse radish peroxidase), which can in turn recognize a biotinylated macromolecule, has been tested in an ELISA system. All derivatives appeared to retain this ability (Figure 2).

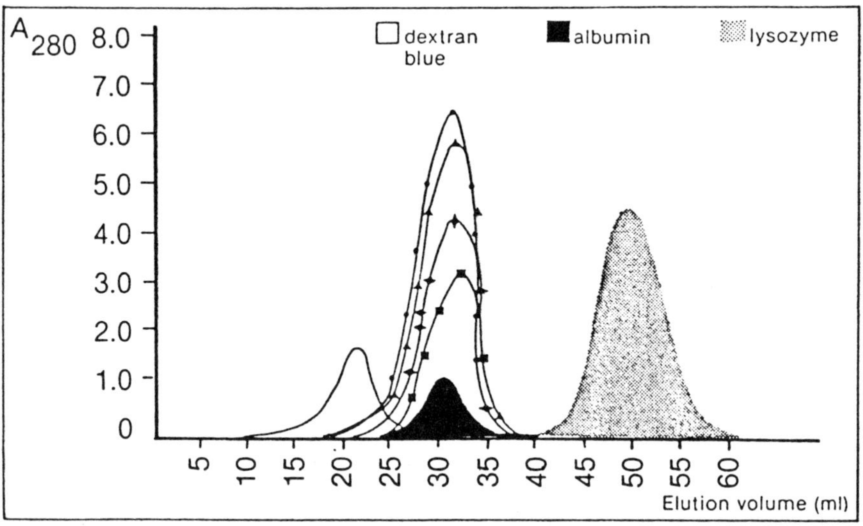

Figure 1. Elution profiles of native (■), acetyl-(◆) succinyl - (▲) and phthalyl- avidins (●) from a Sephadex G-200 column.

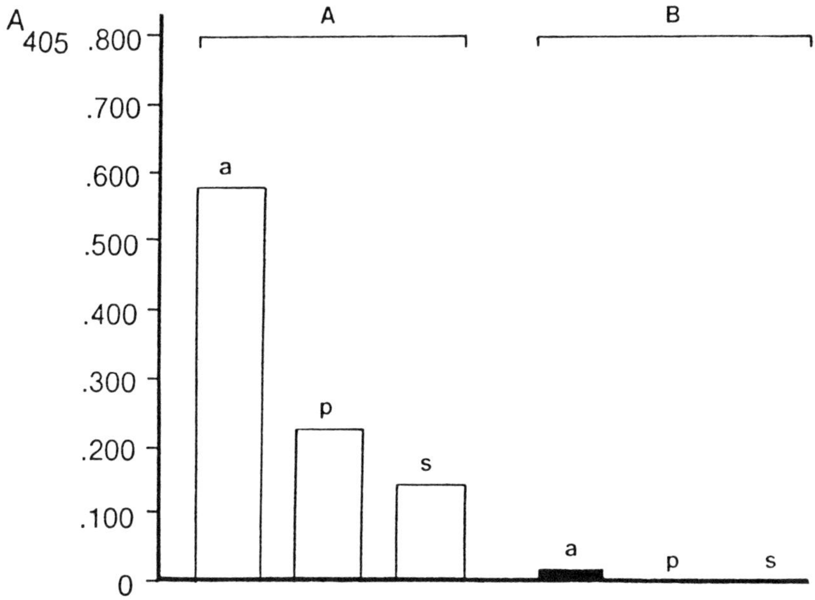

Figure 2. Comparison of the signals obtained with "one step" (A) and "two steps" (B) methods: a = acetyl avidin; p = phthalyl avidin; s = succinyl avidin. The solid phase was prepared by adsorbing a constant amount of a biotinylated antibody to each series of wells of a microtitration plate.

A) "one step" method:

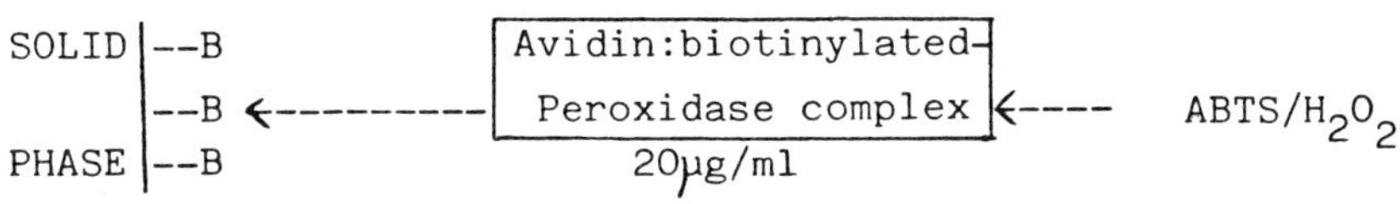

B) "two steps" method:

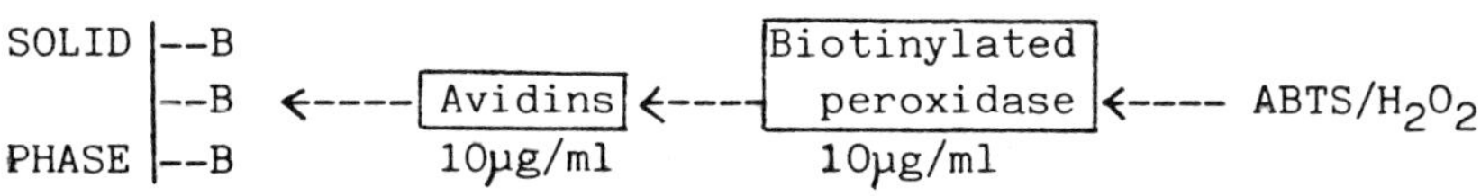

The aspecific binding of complexes of polybiotinylated horse radish peroxidase with native and modified avidins has been tested in different systems (ELISA, immunohistochemistry, DNA blots). Such an aspecific binding was completely abolished in ELISA for all three

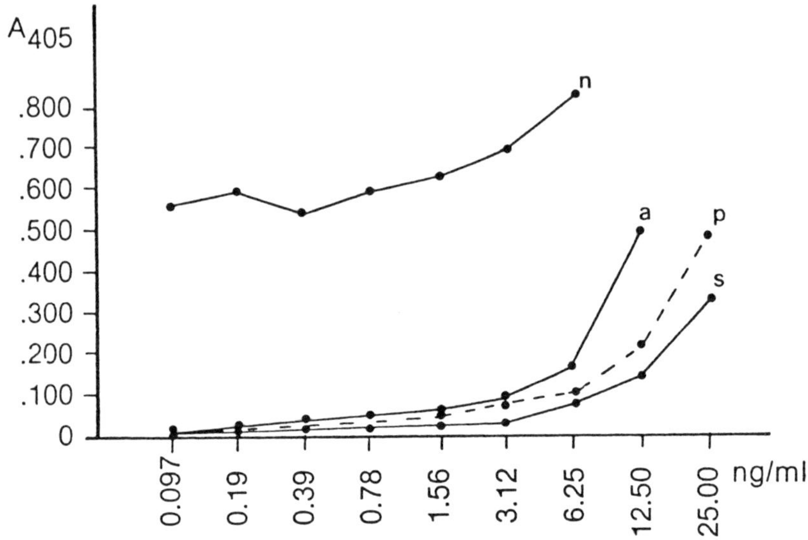

Figure 3. Detection limits of avidin: biotinylated-peroxidase complexes. n=complex with native avidin; a=complex with acetyl avidin; p=complex with phthalyl avidin; s=complex with succinyl avidin. The solid phase was prepared by adsorbing 100 µl of increasing dilutions of a biotinylated monoclonal antibody to each series of wells of a microtitration plate. The antibody concentration was kept constant by the addition of the relevant amounts of the same unbiotinylated antibody. The detection limits for the different complexes in the ELISA system were less than 100 pg, 200 pg and 300 pg of a biotinylated antibody for acetylated, phthalylated and succinylated derivates respectively.

derivatives (Figure 3). In immunohistochemistry and DNA blotting the aspecific binding was completely abolished for the succinylated and phthalylated avidins, while the acetylated derivative showed a very marked decrease, but not total abolition.

Acknowledgements. This work was partially supported by the Progetto Finalizzato CNR "Ingegneria genetica e basi molecolari delle malattie ereditarie."

NEW BIOSPECIFIC CHROMATOGRAPHIES OF β-HEXOSAMINIDASES

C. Emiliani, T. Beccari, J.L. Stirling, R. Sciarra and A. Orlacchio

Istituto Interfacoltà di Chimica Biologica Univ. di Perugia, Italy and Dept. of Biochemistry - King's College, Univ. of London, U.K.

β-N-acetylhexosamidase is one of a group of lysosomal glycosidases whose function is to remove non-reducing monosaccharides from glycolipids, proteoglycans and glycoproteins. In humans, β-N-acetylhexosaminidase exists in two major forms, A and B (1). The A form is the product of two genes and contains α and β subunits while B is composed of β subunits alone (2, 3). The presence of the α subunit confers the ability to hydrolyze the terminal N-acetylglucosamine from the GM2 ganglioside (4) and from the synthetic substrate 4-methylumbellyferyl- β-N-acetylglucosamine-6-sulphate (5). Hydrolysis of the unsulphated substrate, 4-methylumbellyferyl- β-N-acetylglucosaminide and of a variety of other synthetic β-N-acetylglucosaminidases and galattosaminidases are catalyzed equally well by both β-N-acetylhexosaminidases A and B.

The enzyme has gained considerable medical interest since β-N-acetylhexosaminidase A was found to be missing in Tay-Sachs's disease (6), and both A and B in Sandhoff's disease (7).

In addition to the major forms A and B, there are other forms of the enzyme which may be present in small quantities in normal tissues and body fluids, but are sometimes elevated massively in patients with a variety of diseases. β-N-acetylhexosaminidase P has charge properties intermediate between those of A and B. This form increases throughout pregnancy and is the predominant form in maternal parturition blood (8).

A form with similar "intermediate" properties has been described in lymphoblasts from patients with common acute lymphoblastic leukemia (cALL) (9). Using chromatofocusing, Orlacchio et al. (10) have shown that the "intermediate" form occurs not only in cALL cells, but in all leukemic cells isolated from patients or cultured *in vitro* (11).

All the isoenzymes described so far are glycoproteins judging from their behavior on Concanavalin A.

There have been numerous attempts to purify β-N-acetylhexosaminidase isoenzymes using both conventional and biospecific affinity separation techniques. These biospecific methods are of two main types:

a. methods based on specific interactions involving carbohydrate residues carried by the enzymes;

b. methods based on specific interactions involving polypeptides.

a: Lectins immobilized on solid supports were used (12) as a step in the purification of hexosaminidases, but in general they have not proved useful for the separation of individual isoenzymes. However, it is interesting to note that Moss and Besley (13) found differences in the binding of precursor and mature forms of hexosaminidases to lectin columns that possibly indicate differences in glycosylation. The interaction between glycosylated proteins and phenyl boronate agarose (PBA) has proved to be useful in the separation of glycosylated and non-glycosylated forms of hemoglobin (14). Annelid β-N-acetylhexosaminidase isoenzymes can be separated by their interaction with a PBA column (15). This technique has proved useful in the separation of isoenzymes from leukemic cells (16).

b: Specific separations of β-N-acetylhexosaminidases A and B are possible using antisera directed against the α and β subunits. An elegant demonstration of such specificities is to be found in the studies of the maturation of the β-N-acetylhexosaminidase subunits by Hasilik and Neufeld (17). In principle, immunoaffinity columns could be used to purify the enzymes, but as yet the problems of desorption under conditions that yield native enzyme have not been resolved. This constraint has limited the use of immunoaffinity chromatography in the purification of the enzymes.

Small ligands immobilized on solid supports have been used successfully in the purification of β-N-acetylhexosaminidase by affinity chromatography (18-21). In many of these cases, the extent of purification was limited by the use of non-specific methods of elution but nevertheless the techniques used in conjunction with conventional purification procedures have yielded pure enzymes.

Here we illustrate the application of two "biospecific" chromatographic methods to the separation and purification of hexosaminidase isoenzymes, and evaluate the nature of the interaction between the enzyme and the column matrices.

PHENYL BORONATE AGAROSE CHROMATOGRAPHY ON PBA 30 MATREX GEL

In the example we describe here, extracts of the spleen from a patient with Hairy Cell Leukemia were chromatographed onto a column of PBA 30 under the conditions described in Figure 1. Three peaks of enzyme activi-

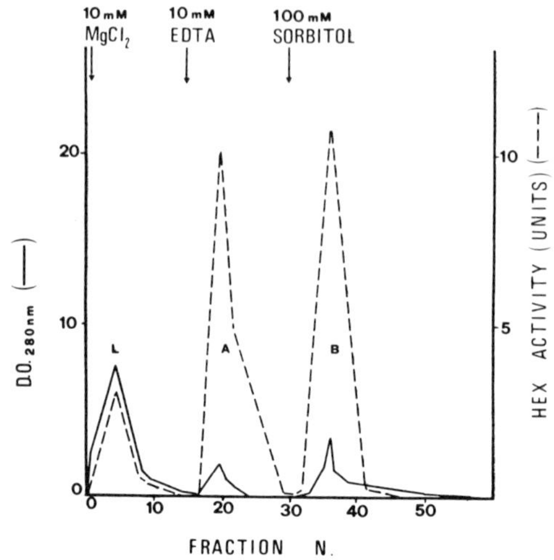

Figure 1. Fractionation of β-N-acetylhexosaminidase isoenzymes from human spleen of a patient with Hairy Cell Leukemia on Matrex Gel PBA-30.

The 40%-70% saturation $(NH_4)_2SO_4$ enzyme preparation was dialysed vs. 20 mM Tris-HCl pH 7.4 containing 1 mM $MgCl_2$, $MnCl_2$ and $CaCl_2$ and pumped through a Concanavalin-A Sepharose 4B column (1 x 25 cm) equilibrated with the same buffer. Then the enzyme was eluted, as a single peak, with a linear gradient of α-methylmannoside in 50 mM Tris-HCl, pH 7.4 containing 1 M NaCl. The active fraction from Concanavalin-A Sepharose was concentrated and dialyzed against 50 mM HEPES pH 8.5 containing 10 mM $MgCl_2$ and applied to a Matrex gel PBA-30 column (0.5 x 6 cm) equilibrated with the same buffer (3 mg prot/ml gel). Adter 30 min pre-incubation, the column was developed with the same buffer at 60 cm/hour and 4 ml fractions were collected until the A280 reached zero. The elution was continued with 50 mM HEPES, pH 8.5, containing 100 mM EDTA and with 50 mM HEPES, pH 8.5, containing 100 mM Sorbitol.

ty were eluted. The first of these, representing 11% of the applied activity, is a form of the enzyme that is lacking in extracts of normal tissues but which occurs in all fresh and cultured leukemic cells. The second and third peaks account for 43% and 26% of the applied activity, respectively. These peaks are observed in all samples of human tissue analyzed, although their proportions are considerably different in, for example, myeloid and lymphoid leukemic cell lines and normal granulocytes and lymphocytes (16).

We were interested to see how the isoenzymes of β-N-acetylhexosaminidase were distributed in the three peaks diluted from PBA 30 and applied each of the peaks, after dialysis, to chromatofocusing on a column of PBE-94. These columns were developed with Polybuffer 74 and the eluates assayed for hexosaminidase activity. The results are shown in Figure 2. In the top profile, a post-Concanavalin A extract of spleen from a patient with Hairy Cell Leukemia was run. The first peak eluted from the column is hexosaminidase B and the second, broad peak eluting after 70 min is hexosaminidase A. The third peak has an isoelectric point lower than pH 4 and is recovered from the column by elution with 1 M NaCl. The second profile in Figure 2 is that given by the first peak eluted from the PBA column. This corresponds to the most acidic component of the total extract and is designated form L because its occurrence was first noted in extracts of leukemic cells. The other two profiles in Figure 2 are for the second and third peaks eluted from the PBA-30 column and clearly correspond to hexosaminidase A and B, respectively. The most striking feature of these results is that the peaks eluted from PBA 30 are eluted in increasing order of their isoelectric points. This is to be expected if there is a predominantly ionic interaction between the negatively charged column and the isoenzymes.

However, the behavior of the isoenzymes is not entire

ly consistent with simple ionic interactions alone. Increasing the EDTA concentration, used to elute hexosaminidase A, to 100 mM did not displace hexosaminidase B, but elution with 100 mM sorbitol gave complete recovery of this isoenzyme. At present we are unable to evaluate the part played by ionic interactions on one hand, and s reversible hydroxyl exchange reactions between cis-diols and the tetrahedral form of boronic acid on the other.

AFFINITY CHROMATOGRAPHY OF HUMAN EPIDIDYMAL ß- N-ACETYL-HEXOSAMINIDASE ON 2-ACETAMIDO-N-(-6-AMINO HEXANOYL)-2-DEOXY-ß-D-GLUCOPYRANOSYLAMINE COUPLED TO CNBr SEPHAROSE 4B (ANAG-SEPHAROSE).

The use of ANAG-Sepharose for the purification of β-N-acetylhexosaminidase is illustrated by recent esperiments with enzymes from human epididymia, Figure 3. We have also used the affinity gel for the purification of hexosaminidases from human liver and placenta (22). By adjusting the concentration of salt in the eluting buffers, it is possible to establish conditions under which first hexosaminidase B and then hexosaminidase A are eluted specifically with N-acetylhexosaminidase.

We consider that this behavior of the enzymes on ANAG-Sepharose has two main components, a specific affinity component on one hand, and non-specific ionic interactions on the other. It is now well known (22) that a primary aliphatic amino group, such as we have on our ligand, reacts with CNBr Sepharose to give an isourea linkage that is protonated at neutral pH. With buffers of low ionic strength, the predominant interaction between the enzymes and the column is ionic and both enzymes can be eluted, first B and then A, by the application of a salt gradient.

In order to exploit the affinity component of the interaction, the salt concentration has to be adjusted to a value just below that required to elute the

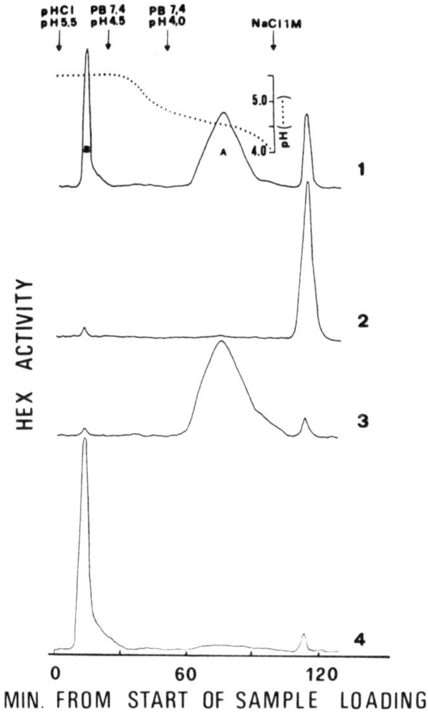

Figure 2. Recording of hexosaminidase isoenzymes from human spleen of a patient with Hairy Cell Leukemia by chromatofocusing on PBE94 coupled with automated enzyme assay.

1. isoenzyme profile of a post-ConA Sepharose extract; 2. profile given by the first peak eluted from a PBA-30 column (isoenzyme L); 3. profile given by the second peak eluted from the PBA-30 (isoenzyme A); 4. profile given by the third peak eluted from the PBA-30 (isoenzyme B). The column was equilibrated with 25 mM piperazine-HCl pH 5.5; the sample was eluted with the above buffer giving peak B. 20 min after sample injection, the elution continued with polybuffer 74 (PB 74), pH 4.9, and 40 min after sample injection with PB 74, pH 4.0. After the gradient, the column was washed with 1 M NaCl and a hexosaminidase activity peak was recovered for the post Con-A extract and isoenzyme L (profiles 1 and 2).

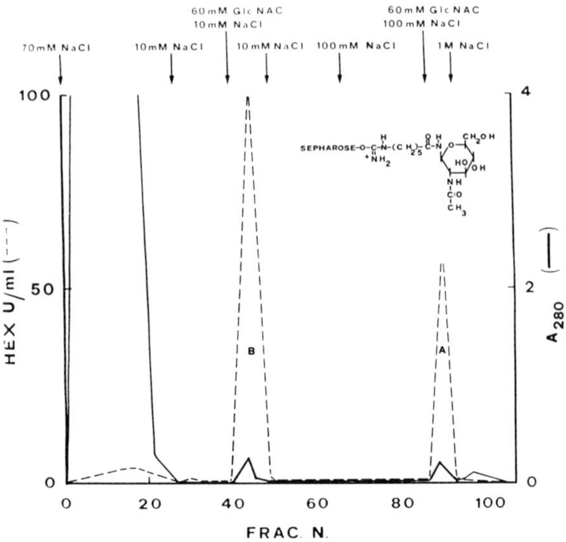

Figure 3. Elution profile of human epididymal β-N-acetylhexosaminidase from ANAG-Sepharose affinity chromatography. The affinity column was prepared by coupling 2-acetamido -N-(6-aminohexanoyl)-2-deoxy- β-D-glycopyranosilamine to CNBr activated Sepharose 4B. The 20-70% satn. $(NH_4)_2SO_4$ enzyme preparation was adjusted to 70 mM NaCl and applied to the affinity column at 60 ml/h. β-N-acetylhexosaminidase B was eluted with 60 mM N-acetylglycosamine in 10 mM NaCl, and form A with the same concentration of N-acetylglucosamine, but in buffer containing 100 mM NaCl. Residual proteins were eluted with 1.0 M NaCl in phosphate buffer.

particular isoenzyme from the column. Under these conditions desorption of the isoenzyme is possible with competitive inhibitors of the enzyme, but not with sugars of related configuration that are non-inhibitors. The existence of a non-specific interaction between hexosaminidase and the ligand was shown using a similar column in which the ligand was changed to the corresponding glucoside, or in which the C-6 spacer arm was blocked with ethanolamine. In neither of these cases was specific desorption of the enzymes possible, although they functioned well as ion-exchange columns. Judging from Ki values for the inhibition of hexosaminidase by free ligand (18 mM) and the ligand concentration in the gel (1 μmole/ml of gel), we would expect very little retardation of the enzymes on the column by the affinity action alone. The fact that we have been able to obtain pure enzyme through biospecific desorption of bound enzyme from the affinity gel (23) is consistent with a mechanism in which weak affinity binding is stabilized by ionic interactions between the enzymes and the cationic groups on the gel.

Although the use of soft affinity gels such as ANAG-Sepharose can yield useful results, the separations are still relatively slow. We can purify milligram quantities of enzymes starting from a crude $(NH_4)_2SO_4$ fraction in about two days. Much of this time is spent in reducing the non-specific protein background to acceptably low levels before specific elution of the hexosaminidases. Inevitably, a loss of enzyme activity results from such protracted elution protocols.

High performance affinity chromatography promises to increase the speed with which affinity separations can be achieved. In preliminary experiments we have coupled the ligand p-aminophenyl-2-acetamido-2-deoxy-1-thio-β-D-glucopyranoside to a Beckman Ultraffinity EP column (0.5 ml). A matrix made in this way should be uncharged and thus relatively free from unwanted non-specific ionic effects. The loss of charge stabilization of the affinity interaction between the enzyme and ligand that we exploit in ANAG-Sepharose should be offset by a potentially much higher ligand concentration on the ultraffinity column.

A high proportion of human placental β-N-acetylhexosaminidase (0.1 unit applied, 90% retained) was retained by the column when the sample was loaded in 10 mM sodium phosphate buffer, pH 6.0 containing 50 mM Na_2SO_4. Hexosaminidase A was eluted from the column with 10 mM sodium phosphate buffer pH 6.0 containing 0.3 M NaCl. Elution of hexosaminidase B proved to be difficult and was only possible when 10 mM NaCl and 80 mM N-acetylhexosaminidase were used in combination. Overall, the specific activity of the isoenzymes was increased one thousand fold, but only 25% of the applied activity was recovered.

Work is now in progress to assay the effect of using the ANAG ligand on the column and to develop the technique so that it can be used for the preparation of milligram quantities of the enzymes.

Acknowledgements. Work was supported by MPI and CNR. Ultraffinity EP columns were generously provided by Beckman Analytical SpA, Milan, Italy; PBA 30 Matrex gel from Amicon, Grace Italiana SpA, Milan. Carla Emiliani, Ph.D. is an AIRC Fellow. We thank Mr. D. Piccini for his technical assistance.

REFERENCES

1. Robinson, D. and Stirling, J.L. (1968) Biochem. J. 321-327.
2. Lalley, P.A., Rattazzi, M.C. and Shows, T.B. (1974) Proc. Natl. Acad. Sci. USA 71, 1569-1573.
3. Gilbert, F., Kucherlapati, R., Creagan, R.T., Murname, M.J., Darlington, G.J. and Ruddle, F.H.(1975) Proc. natl. Acad. Sci. USA 71, 263-267.
4. Leaback, D.H. and Walcher, P.G. (1961) Biochem. J. 78, 151-156.
5. Beccari, T., Emiliani, C. and Orlacchio, A., Hosseini, R. and Stirling, J.L. (unpublished results).
6. Okada, S. and O'Brien, J.S. (1969) Science 165, 698-700.
7. Sandhoff, K.(1969) FEBS Lett. 4, 351-354.
8. Stirling, J.L. (1972) Biochim. Biophys. Acta 271, 154-162.

9. Ellis, R.B., Rapson, N.T., Patrick, A.D., Greaves, M. F. and Path, M.R.C. (1978) New Engl. J. Med. 298, 476-480.
10. Orlacchio, A., Emiliani, C., Rambotti, P., Pioda, G.B. and Davis, S. (1986) Biochem. Med. and Met. Biol., in press.
11. Orlacchio, A., Emiliani, C., Tabilio, A. and Pioda, G. B. (1986) Cell. Biochem. Funct. 4, 197-203.
12. Brattain, M.G., Kimball, P.M., Pretlow, T.G. and Marks, M.E. (1977) Biochem. J. 163, 247-251.
13. Moss, S.E. and Besley, G.T.B. (1985) Glycoconjugate 2, 267-277.
14. Day, S., Thornburg, R., Thorpe, S. and Baynes, J. (1979) J. Biol. Chem. 254, 9304-9398.
15. Orlacchio, A., Emiliani, C., Maffei, C. and Beccari, T. (1985) Cell. Mol. Biol. 31, 163-169.
16. Orlacchio, A., Maffei, C., Emiliani, C., Rambotti, P. and Davis, S. (1984) Biochem. Biophys. Res. Commun. 122, 966-973.
17. Hasilik, A. and Neufeld, E.F. (1980) J. Biol. Chem. 255, 4937-4945.
18. Grebner, E.G., and Parikh, I. (1974) Biochim. Biophys. Acta 350, 437-441.
19. Geiger, B., Ben-Yoseph, Y. and Aron, R. (1974) FEBS Lett. 45, 276-280.
20. Bedi, G.S., Shah, R.H. and Bahl, O.M.P. (1984) Archives Biochem. Biophys. 233, 237-250.
21. Orlacchio, A., Emiliani, C., Di Renzo, G.C. and Cosmi, E.V. (1986) Clin. Chim. Acta, in press.
22. Nishikawa, A.H. and Bailon, P. (1975) Archives Biochem. Biophys. 154, 576-589.
23. Dewji, N.N., De Keyzer, D.R. and Stirling, J.L. (1986) Biochem. J. 234, 157-162.

PREPARATIVE APPLICATIONS OF SUBUNIT EXCHANGE CHROMATOGRAPHY

Emilia Chiancone and Maurizio Gattoni

CNR Center of Molecular Biology, Department of Biochemical Sciences, University 'La Sapienza', 00185 Rome, Italy

In the purification of macromolecules bioaffinity methods are the procedures of choice since they enable one to obtain high purity products in a significantly reduced number of steps. Subunit exchange chromatography is such a method that permits the purification of polymeric and self-associating systems. It exploits the specificity of subunit recognition and in particular the reversible interactions of matrix-bound protein subunits with soluble subunits of the same or closely related proteins (1-6).

The extent of subunit association can be controlled by such factors as pH, composition and ionic strength of the medium, presence of cofactors. This immediately suggests a simple two step procedure using immobilized subunits to extract a polymerizing system from a multicomponent mixture. In the first step, carried out under conditions that favor subunit association, the interaction between immobilized and soluble subunits leads to an exchange of subunits between the liquid and the solid phase and part of the protein that was initially in solution is bound to the matrix in a specific way. In the second step the matrix-bound polymer thus formed is eluted under conditions that promote dissociation into subunits, regenerating also the immobilized subunits, ready for a new purification cycle after equilibration with the associating buffer.

This simple procedure, relying on knowledge of conditions that affect the association-dissociation behavior of the system, has proved effective for purification of a variety of self-associating and oligomeric systems from tissue homogenates and biological fluids. Immobilized subunits of the same or of a homologous protein capable of forming hybrids with the system to be isolated have been employed (2, 7-15). These studies and detailed frontal analyses of analytical subunit exchange chromatography experiments have shown that immobilized subunits retain essentially the same associating capacity as the protein in solution provided multipoint attachment and/or coupling of the protein to the matrix via more than one subunit are minimal (2-4, 7-15).

The present paper provides the guidelines for the preparation of properly immobilized subunits and for the measurement of their associating capacity. In addition the use of subunit exchange chromatography is exemplified by its application to the extracellular α-amylase of Bacillus subtilis, an organism that has been reclassified recently as B. amyloliquefaciens (16). This bacterial α-amylase undergoes a reversible association to dimers that depends on the presence of zinc ions which stabilize the dimeric form of the enzyme (17).

EXPERIMENTAL

I. Preparation of Immobilized Subunits

Immobilized subunits are best obtained by coupling the protein to the activated matrix. Sepharose 4B and the standard CNBr activation method have been employed extensively (7, 9-11, 13-15). One drawback of the CNBr activation is its lack of chemical selectivity, which may lead to heterogeneity of the immobilized subunits. However, by varying the pH of the reaction, preferential binding to the matrix via the α-or ε-aminogroups may be achieved. In any case it is advisable to use low CNBr and protein concentrations (1-50 mg/ml of settled gel and 1-10 mg/ml, respectively) in order to minimize multipoint attachment to the matrix.

During the immobilization of oligomeric proteins special precautions should be taken to avoid coupling of the subunits through residues that lie at or near relevant intersubunit contact regions. To this end the oligomeric protein should be immobilized under conditions that stabilize the oligomer (18). Further precautions to reduce steric hindrance effects may be: reduction of the concentration of activated groups, which can be achieved by washing the matrix with 0.1 M sodium bicarbonate (15); ethanolamine present during the coupling step, roughly equimolar to the potentially reactive groups on the protein (18); indirect coupling of the subunits to the matrix via an appropriate spacer arm (8).

At the end of the coupling reaction any non-covalently-bound protein has to be removed. Hence the matrix is washed extensively under conditions favoring dissociation of the protein and immobilized subunits are generated. In the case of oligomeric proteins, studies to date have shown that the average size of the matrix-bound material varies even when the coupling reaction has been carried out under conditions that stabilize the oligomer. Thus, proteins may be immobilized as oligomers, as subunits of different states of association or as monomers (1, 8, 11, 15, 18, 19).

2. Measurement of the Associating Capacity of the Immobilized Subunits

The associating capacity of the immobilized subunits, and hence the amount of matrix-bound polymer formed under given experimental conditions, is governed by the law of mass action and depends on several parameters such as the concentrations of immobilized and soluble protein, the association constants in solution and in the solid phase and the stoichiometry of association of the system (1, 4).

An easy means to assess the capacity of a preparation of immobilized subunits to interact with soluble ones involves the use of a small chromatographic column (5–8 ml of settled immobilized subunits) equilibrated with associating buffer. For a qualitative indication of the associating capacity it suffices to percolate a small volume (~1 ml) of soluble protein in the same associating buffer through the column. If the soluble protein interacts with the immobilized subunits, its elution volume will be larger than that of a non-interacting protein, i.e. than the void volume of the column, V_o (see Fig. 1a). For a quantitative measure of the associating capacity, soluble protein should be applied to the column until a steady state is reached, that is when the concentrations of the effluent and of the inflowing solution are equal. The amount of soluble protein retained specifically on the column, $\bar{Y}$, is then given by the product of the protein concentration in solution, c, and the increase in elution volume with respect to the void volume, $\Delta V = V - V_o$ (see Fig. 1b).

If desired the experimental values of $\bar{Y}$ can be compared with those expected on the basis of the law of mass action for a given set of association constants in solution and in the solid phase. For a monomer-dimer equilibrium, assuming the rate of equilibration fast in compared with the rate of flow through the column, the value of $\bar{Y}$ is related to the amount of monomer covalently bound to the matrix, $R\text{-}X_t$, and to the association constants in solution and in the solid phase, respectively L_1 and L_2 (in dl/g), by the following expression:

$$\bar{Y} = \frac{(R\text{-}X_t) / 2}{\dfrac{L_1}{L_2} \; \dfrac{(X)}{(X_2)}}$$

The ratio $(X)/(X_2)$ here corresponds to the ratio $\alpha/(1-\alpha)$

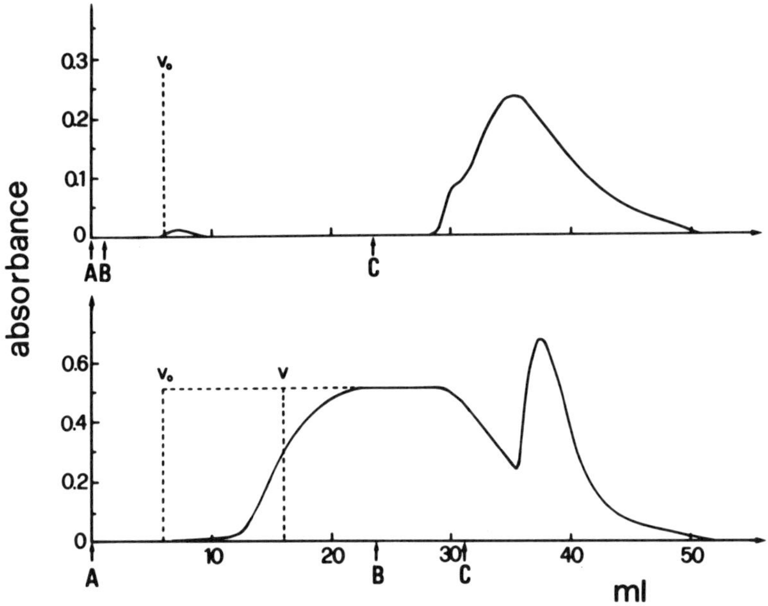

Fig.1. Elution profiles of an α-amylase solution from a column containing Sepharose 4B-bound α-amylase. (a) pulse and (b) column saturation experiment. The column (5 ml of settled gel) was equilibrated with associating buffer (0.1 M NaCl, 0.005 M calcium acetate, 50 μM zinc acetate at pH 7.0) at 8°C. Flow rate, 11 ml/h. The protein concentration in the solid phase was 4.6 mg/ml. Arrows indicate: A, application of an α-amylase solution at 2 mg/ml; B, application of associating buffer; C, application of dissociating buffer (0.01 M Na-EDTA at pH 7.0). V_o and V represent the elution volume of an inert protein and of α-amylase, respectively. In (b) the area $(V - V_o) \times c$ corresponds to the amount of soluble α-amylase bound specifically to the column at the steady state concentration, c.

for the protein in solution, α being the degree of dissociation. α is a function of the total concentration in solution at equilibrium, c, according to the equation

$(1 - \alpha)/\alpha^2 = L_1 c$. $(R\text{-}X_t)$ is obtained by multiplying the volume of the column by the concentration of the covalently-bound monomer. The latter can be determined either spectrophotometrically after minimizing the turbidity of the material with a protein-free gel in the reference beam (1, 2, 18) or by aminoacid analysis after hydrolysis of measured gel aliquots with 6 N HCl for 24 h at 110°C (3).

From the relationships just given one can calculate the dependence of the elution volume on the protein concentration in solution. As shown in Fig. 2 the elution volume is inversely related to the concentration of soluble protein, i.e. the elution volume of the interacting protein is largest when its concentration in solution is low. This property is very useful when the immobilized subunits are used in columns during protein purification; alternatively immobilized subunits can be employed in batch procedures or in closed reactors.

3. Application of Subunit Exchange Chromatography to Bacterial α-Amylase.

The extracellular α-amylase of B. subtilis is a metalloprotein that contains 3 or 4 gatoms of calcium and a half gatom of zinc per monomeric unit of the enzyme (M_r 48000). Detailed ultracentrifuge, light scattering and gel filtration studies (17, 20, 21) have shown that this bacterial amylase undergoes a dimerization reaction which depends on pH and, more importantly, on the presence of zinc ions. Thus, at neutral pH values α-amylase (Zn^{2+}-free) has a fairly low tendency to associate into dimers (L_1 = 1 dl/g), but the presence of Zn^{2+} displaces the equilibrium in the direction of dimers because of its preferential binding to this species (17). In a typical purification procedure of this α-amylase, which involves an initial ammonium sulfate precipitation from the culture supernatant and three chromatographic steps (on DEAE- and CM-cellulose and on hydroxylapatite), a major problem is constituted by the copurification of active alkaline and neutral protease exoenzymes (22). This would be avoided by a bioaffinity

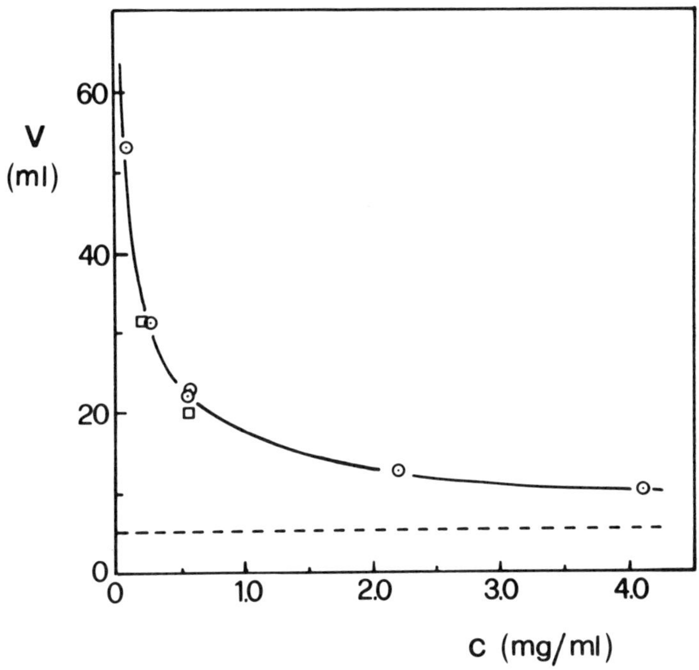

Fig. 2. Elution volume of an α-amylase solution from a column containing Sepharose 4B-bound α-amylase as a function of protein concentration in solution. Buffer: 0.1 M NaCl, 0.005 M calcium acetate, 50 μM zinc acetate at pH 7.0. The protein concentration in the solid phase was 4.6 mg/ml of settled gel, the bed volume was 5 ml and the flow rate 11 ml/h. Temperature: 8°C (○), 20°C (□). The dashed line represents the elution volume of an inert protein; the continuous line is a theoretical one calculated with the equations given in the text.

method for specifically extracting α-amylase from the culture supernatant or the ammonium sulfate precipitate.

In order to investigate the applicability of subunit exchange chromatography, α-amylase (bacterial α-amylase type II from Sigma Chemical Co.) was immobilized on CNBr activated Sepharose 4B by using 60 - 100 mg of CNBr and 7 - 10 mg of protein per ml of settled gel. The coupling step was carried out for 24 hours in the cold, under conditions that stabilize the dimer i.e. in 0.1 M NaCl, 0.005 M calcium acetate and 50 μM zinc acetate at pH 7.0 (17, 23). The unreacted groups were quenched by washing the resin extensively with the coupling buffer; thereafter any non-covalently-bound protein was removed with a dissociating solution (0.01 M sodium EDTA at pH 7.0). The amount of α-amylase immobilized on the matrix was determined spectrophotometrically at 280 nm on the basis of the extinction coefficient $E^{1\%}_{1\,cm}$ = 25.6 (20) using a protein-free gel in the reference beam to minimize the turbidity of the material.

The α-amylase monomers thus immobilized interact with the protein in solution. This is immediately apparent from experiments as discussed above employing a small column of immobilized enzyme (~5 ml) equilibrated with associating buffer (0.1 M NaCl, 0.005 M calcium acetate and 50 μM zinc acetate at pH 7.0). When a small pulse of α-amylase solution at a concentration of 2 mg/ml in the same buffer is applied to the column, its elution is retarded significantly with respect to the void volume and the protein is eluted after application of the dissociating buffer (Fig. 1a). If the column is saturated with the α-amylase solution, it can be estimated that under these conditions the specifically bound protein amounts to 14 mg/20 mg of settled immobilized subunits (Fig. 1b).

A quantitative study was then performed in which a column of immobilized enzyme was saturated with α-amylase solutions of varying concentrations. Figs.2 and 3 show, respectively, the elution volume, V, and the amount of protein retained on the column, $\bar{Y}$, plotted as a function of the α-amylase concentration in solution. In particular Fig. 2 shows that the elution volume is five or ten times the void volume when the enzyme concentration in solution is of the order of 0.1 - 0.25 mg/ml; under these conditions the enzyme could easily be separated from any contaminating protease in a single step. On the other hand Fig. 3 brings

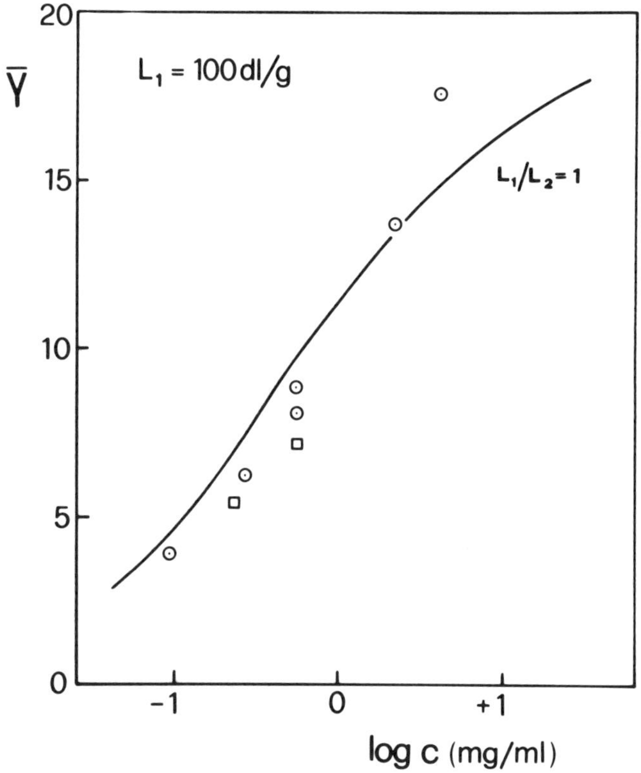

Fig. 3. Amount of α-amylase retained on a column containing Sepharose-4B bound α-amylase as a function of protein concentration in solution. The experimental points are taken from Fig. 2 and refer to α-amylase in 0.1 M NaCl, 0.005 M calcium acetate, 50 μM zinc acetate at pH 7.0 and at a temperature of 8 (○) and 20°C (□).

out that the data conform to a monomer-dimer equilibrium with association constants in solution (L_1) and in the solid phase (L_2) of 100 dl/g. Figs. 2 and 3 also show that between 8 and 20°C temperature hardly affects the associating capacity of immobilized α-amylase.

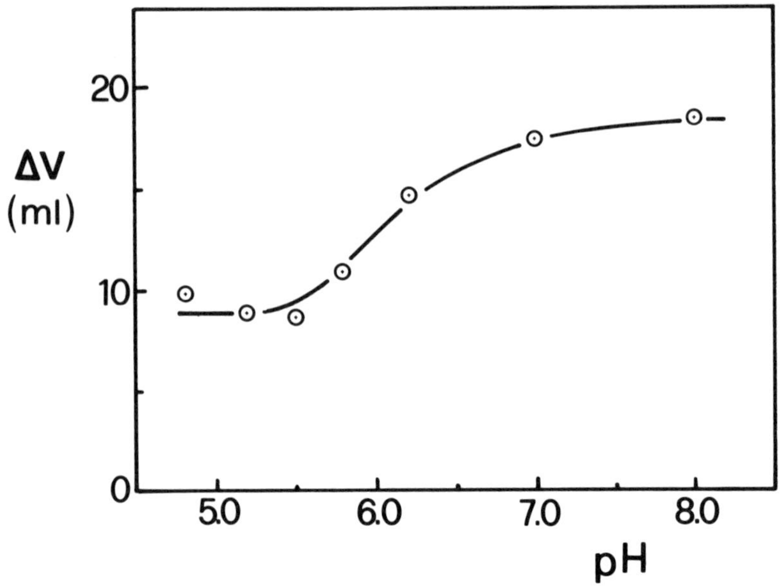

Fig. 4. Effect of pH on the elution volume of α-amylase from a column containing Sepharose 4B-bound α-amylase. The α-amylase solution was at a concentration of 0.5 mg/ml. Temperature, 8°C. The column and the experimental conditions are the same as in Fig. 2.

The effect of pH was analyzed in a further set of the same type of experiments. Solution studies have shown that dimer formation diminishes when the pH is lowered from 7 to 5, a phenomenon that has been ascribed to the involvement of histidyl residues in the chelation of zinc ions (23). Over this same pH range the interaction between immobilized and soluble α-amylase decreases as depicted in Fig. 4.

CONCLUSIONS

Self-associating and oligomeric proteins may be purified in a simple and effective way by means of subunit exchange chromatography. Major advantages of the method are the very low aspecific binding of extraneous proteins (at ionic strengths $\geq$ 0.01 M) and the fact that subunits from a closely related protein, capable of forming hybrids with the system to be isolated, can be immobilized.

The difficulties that may arise due to heterogeneity in the covalent attachment of the subunits to the matrix and/or to their coupling via residues near the intersubunit contact regions in general may be solved by appropriate changes in the coupling conditions.

REFERENCES

1. Antonini, E., Rossi Fanelli, M.R. and Chiancone, E. (1975) in: Protein-Ligand Interactions (Sund, H., ed.) pp. 45-59, Walter de Gruyter, Berlin.
2. Antonini, E. and Rossi Fanelli, M.R. (1976) Methods Enzymol. 44, 538-546.
3. Antonini, E., Carrea, G., Cremonesi, P., Pasta, P., Rossi Fanelli, M.R. and Chiancone, E. (1979) Anal. Biochem. 95, 89-96.
4. Chiancone, E., Gattoni, M. and Antonini, E. (1983) in: Affinity Chromatography and Biological Recognition (Chaiken, I.M., Wilchek, M. and Parikh, I., eds.) pp. 103-112, Academic Press, Orlando, FL.
5. Chiancone, E. and Gattoni, M. (1986) J. Chromatography 376, 343-348.
6. Chiancone, E. and Gattoni, M. Methods Enzymol., 135, in press.
7. Van Etten, R.L. and Waheed, A. (1980) Arch. Biochem. Biophys. 202, 366-373.
8. Takahashi, M. and Gross, E.L. (1978) Biochemistry 17, 806-810.
9. Carrea, G., Pasta, P. and Antonini, E. (1983) Biotechnol. Bioeng. 25, 1331-1339.
10. Carrea, G., Pasta, P. and Antonini, E. (1985) Biotechnol. Bioeng. 27, 704-710.
11. Muronets, V.I., Ashmarina, L.I. and Nagradova, N.K. (1981) Experientia 37, 15-16.

12. Bode, J. and Wagner, K.G. (1975) Biochem. Biophys. Res. Commun. 62, 868-876.
13. Carrea, G., Lugaro, G., Niada, R., Vecchini, P. and Antonini, E. (1979) FEBS Lett. 104, 393-395.
14. Shuder, S., Wittenberg, J.B., Haseltine, B. and Wittenberg, B.A. (1979) Anal. Biochem. 92, 473-481.
15. Gurne, D., Chen, J. and Shemin, D. (1977) Proc. Natl. Acad. U.S.A. 74, 1383-1387.
16. Welter, N.E. and Campbell, L.L. (1967) J. Bacteriol. 94, 1124-1130.
17. Tellam, R. and Winzor, D. (1978) Biochem. J. 173, 185-190.
18. Rossi Fanelli, M.R., Amiconi, G. and Antonini, E. (1978) Eur. J. Biochem. 92, 253-259.
19. Chan, W.W.C. (1970) Biochem. Biophys. Res. Commun. 41, 1198-1204.
20. Kakiuchi, K., Kato, S., Imanishi, A. and Isemura, T. (1964) J. Biochem. (Tokyo) 55, 102-109.
21. Kakiuchi, K., Hamaguchi, K. and Isemura, T. (1965) J. Biochem. (Tokyo) 157, 167-175.
22. Mäntsälä, P. and Zalkin, H. (1979) J. Biol. Chem. 254, 8540-8547.
23. Isemura, T. and Kakiuchi, K. (1962) J. Biochem. (Tokyo) 51, 385-392.

USE OF MONOSPECIFIC ANTIBODIES FOR THE PURIFICATION OF D-AMINO ACID OXIDASES FROM VARIOUS SOURCES

G.L. Fossati*, P. Fusi, E. Monti, B. Curti

Dipartimento di Fisiologia e Biochimica Gen.li,
Università di Milano, Via Celoria 26, 20133
Milan; *Recordati S.p.A., 20148 Milan, Italy

D-amino acid oxidase (E.C.1.4.3.3., DAAO), a flavoprotein which catalyzes the oxidative deamination of D-amino acids, has been purified to homogeneity from hog kidney (1). The physiological role of this enzyme remains obscure, even if alternative hypotheses on its function have been proposed in recent years (2). In spite of its apparent uselessness, D-amino acid oxidase has been largely investigated as a model system for flavo-oxidases, and its kinetic parameters, physical-chemical properties, catalytic mechanism and primary structure have been determined (3,4). Unfortunately, the three-dimensional structure of the enzyme is still unknown due to the difficulty of obtaining crystals of hog enzyme suitable for X-ray studies (5). A solution to this problem would be to search for other sources of the enzyme: such a project would require much screening among various organisms and a relatively rapid purification procedure of this flavo-protein. Bioaffinity methods might represent a useful tool to apply in such a case, provided a suitable recognition system of the enzyme molecule were available.

In the present paper, we report the preparation of monospecific anti-D-amino acid oxidase antibodies from hog kidney, their immobilization on a solid phase and the use of such a system for the rapid purification of D-amino acid oxidases from various sources.

MATERIALS AND METHODS

D-amino acid oxidase from hog kidney was purified to homogeneity according to Curti, et al. (1). Serum rabbit antibodies to the enzyme were prepared as described in Hurn, et al. (6) and purified through ammonium sulphate fractionation and DEAE chromatography (7). In order to obtain monospecific antibodies, D-amino acid oxidase was immobilized on an LKB Ultrogel AcA 22 through the glutaraldehyde procedure (8). The pre-purified anti-DAAO antibodies were found to bind selectively to a column of Ultrogel AcA 22-DAAO (4 x 1 cm). The column was washed first with 20 mM potassium phosphate buffer, pH 7 + 0.15 M NaCl, followed by 100 mM borate buffer pH 8.5 + 0.5 M NaCl (BBS) + 0.1% NP 40. The monospecific anti-DAAO antibodies were finally eluted by 200 mM glycine-HCl buffer pH 2.5 + 10% ethylene glycol. The antibodies were then coupled to Affigel 10 Biorad, according to the manufacturer's instructions, with slight modifications (9). The solid phase coupled antibodies were then used as a final step for the purification of D-amino acid oxidases from different sources.

Enzyme activity was determined spectrophotometrically with D-α-phenylglycine as substrate as described in Ferti, et al. (10), or by measuring the oxygen consumption through a Clark electrode as reported in (10) but with D-proline as substrate. The assay was carried out at 37°C.

Protein determination was performed by the biuret method (11) or by using the Coomassie blue procedure according to Pierce (12). In some cases homogeneous D-amino acid oxidase from hog kidney was used as standard.

SDS-gradient gel electrophoresis was carried out as described by O'Farrell (13). Protein bands were stained with Coomassie blue or by the silver staining method (14).

RESULTS AND DISCUSSION

Our initial screening of alternative sources of DAAO was confined to mammalian tissues , namely calf cerebellum rat liver, and sheep, rat and horse kidney. In all cases the presence of the enzyme was ascertained by assaying the activity on raw tissue homogenates; cross-reactivity of the enzyme species with anti-DAAO antibody from hog kidney was also checked through blot-transfer of SDS-gradient gel electrophoresis analyses on raw extracts of the above tissues. A typical purification procedure, carried out at 4°C unless otherwise stated, consisted of: a) Tissue homogenate. The tissue was homogenized in 16 mM pyrophosphate buffer pH 8.3, in 1:3 ratio, w/v. The homogenate was then centrifuged for 30 min. at 8,000 rpm and the pellet discarded. b) Heat and pH treatment. The supernatant of the previous step was heated to 38°C and maintained at this temperature for 20 min. with continuous stirring. The pH of the homogenate was then adjusted to 5.2 with 2 M acetic acid and the solution centrifuged at 8,000 rpm for 30 min. and the pellet discarded. c) Ammonium sulphate fractionation. The supernatant of the previous step was brought to 42% saturation with ammonium sulphate, and centrifuged at 8,000 rpm for 1 hour. The precipitate was redissolved in 20 mM potassium phosphate buffer pH 7 + 0.15 M NaCl and dialyzed overnight against the same buffer. d) Affinity chromatography. The dialyzed solution was then passed through the Affigel 10 column and recycled through it for about 1 hour with the dialysis buffer, to ensure complete binding of the enzyme. Washings were then carried out with BBS buffer + 0.1% NP 40, followed by 1 M $MgCl_2$, and finally by 2.8 M $MgCl_2$ which selectively eluted the bound enzyme. The column was then washed with 0.2 M glycine-HCl pH 2.5 + 10% ethylene glycol, to remove any traces of residual bound material, and immediately re-equilibrated to neutral pH with the starting phosphate buffer. Previous attempts to elute the enzyme in an active form by 0.2 M glycine-HCl pH 2.5 or through the use of various molarities of KCNS or KJ were unsuccessful.

Table 1 shows the results of a typical purification procedure of D-amino acid oxidase isolated from rat kidney. The whole procedure takes two working days and gives an enzyme with a purification factor of $\simeq$ 800 fold in comparison to the starting material and a final yield of $\simeq$ 40%. On the average, enzyme yields for the other sources investigated were around 22% and purification factors ranged from 700 to over 15,000.

TABLE 1

PURIFICATION OF D-AMINO ACID OXIDASE FROM RAT KIDNEY

Purification steps	Volume (ml)	Activity (U/ml)*	Total units	Protein (mg/ml)	Specific activity (U/mg)	Purification factor	Recovery (%)
a) Homogenate	166.0	5.14	853.2	27.60	0.186	1	100
b) Heat and pH treatment	134.0	4.75	636.5	19.30	0.246	1.32	74.6
c) Ammonium sulphate fraction	3.7	162.00	599.4	49.70	3.250	17.50	70.2
d) Antibody-affinity chromatography	85.4	4.35	371.6	0.03	145.00	779.60	43.5

* One unit of activity corresponds to the consumption of 1 μmole of O_2 per min.

The large scattering in the purification index could partly be ascribed to the differences in specific activities of the homogenates. The purity of the enzyme following the affinity chromatography step was checked by SDS-gradient gel electrophoresis, as shown in Figure 1. In the same figure, the effectiveness of the purification method is also shown by a comparison among protein patterns at various stages of the purification procedure. The rat kidney DAAO as well as the enzyme from other sources showed the same molecular weight of 40 Kd/monomer on SDS-gradient gel electrophoresis, as for the hog kidney D-amino acid oxidase (see Figure 1). Preliminary kinetic data - such as substrate specificity and pH optima - of the species examined again showed no significant differences as compared to the hog enzyme (data not shown).

The procedure, as already mentioned, is quite rapid. The present antibody-affinity column is tailored for 1.5 mg of purified protein, but it could be scaled up for the preparation of larger quantities of the enzyme. Previous methods of purification of this enzyme (1,15,16) are more laborious; furthermore, they require two or three chromatographic steps and are not always applicable to enzyme sources other than hog kidney.

The monospecific antibodies after the Ultrogel AcA 22-DAAO column are free from protein contaminants (see Figure 1). The antibody-affinity column is quite stable: in our experimental conditions, it has been working for over one year at 4°C, with only moderate loss of its binding capacity (≃25%).

Experiments are in progress to determine amino acid composition and peptide mapping of the various DAAO isolated, in order to acquire useful information for crystallization studies of the proteins.

The biorecognition system of a macromolecule based on the use of the respective solid-phase bound antibody seems, through our experience, to be very promising as a "detector" of homologous proteins from different sources. In fact, through the use of immunoscreening-blot transfer procedures our anti-hog DAAO antibodies were able to detect D-amino acid oxidases not only in mammalian tissues, but also in birds, amphibians,

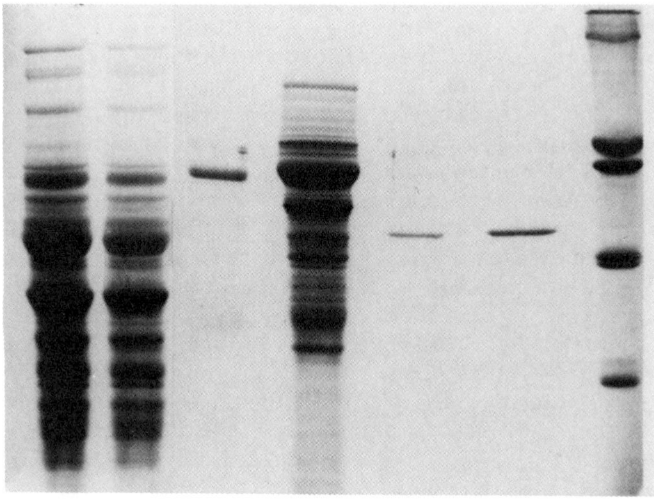

Figure 1. SDS-gradient gel electrophoresis.
Lanes 1, 2, 4, 6 - Rat kidney DAAO purification procedure, steps a), b), c), and d), respectively.
Lane 3 - Anti-hog kidney DAAO antibody, after Ultrogel AcA 22 - DAAO column (monospecific antibodies).
Lane 5 - Homogeneous DAAO from hog kidney.
Lane 7 - Molecular weight markers, from top to bottom: phosphorylase b (MW 94,000), serum albumin (MW 67,000), ovalbumin (MW 45,000), carbonic anhydrase (MW 30,000), trypsin inhibitor (MW 20,100).

fish and yeast (E. Gavazzi, et al., manuscript in preparation). In prokaryotes, D-amino acid oxidase has not been detected at the present time. In this regard, it is worth mentioning that in E. coli cell extracts, our bioaffinity method showed cross-reactivity between the anti-hog kidney DAAO antibody and D-alanine dehydrogenase, a membrane enzyme. The latter, a flavodehydrogenase and not a flavo-oxidase enzyme, utilizes quinones instead of oxygen as the final electron acceptor. Thus, the method seems apt to signal not only the presence of homologous proteins, but also that of different proteins sharing similar structural domains.

Acknowledgments. Work supported by a grant from the Ministero della Pubblica Istruzione, Italy.

REFERENCES

1. Curti, B., Ronchi, S., Branzoli, U., Ferri, G. and Williams, C.H. Jr. (1973) Biochim. Biophys. Acta 327, 266-273.
2. Hamilton, G.A. (1985) Adv. Enzymol. 57, 86-178.
3. Bright, H.J. and Porter, D.J.T. (1975) in: The Enzymes (Boyer, P.D., ed.) pp. 421-505, Academic Press, New York.
4. Ronchi, S., Minchiotti, L., Galliano, M., Curti, B., Swenson, R.P., Williams, C.H. Jr., Massey, V. (1982) J. Biol. Chem. 257, 8824-8834.
5. Bolognesi, M., Ungaretti, L., Curti, B., Ronchi, S., (1978) J. Biol. Chem. 253, 7513-7514.
6. Hurn, B.A.L. and Chantler, S.M. (1980) Methods Enzymol. 70, 104-142.
7. Kaminsky, L.S., Fasco, M.J. and Guengerich, F.P.(1981) Methods Enzymol. 74, 262-272.
8. Lilly, M.D. (1976) Methods Enzymol. 44, 46-53.
9. Biorad Technical Bulletins no. 1085 and 1099, Bio-Rad, Richmond, CA
10. Ferti, C., Curti, B., Simonetta, P.M., Ronchi, S., Galliano, M. and Minchiotti, L. (1981) Eur. J. Biochem. 119, 553-557.
11. Brumby, P.E. and Massey, V. (1968) Biochem. Prep. 12, 29-41.
12. Bradford, M. (1976) Anal. Biochem. 72, 248-254.
13. O'Farrel, P.H. (1975) J. Biol. Chem. 250, 4007-4021.
14. Morrissey, J.H. (1981) Anal. Biochem. 117, 307-310.
15. Jenkins, W.T., Behlen, A. and Rogers, P.J. (1979) Anal. Biochem. 94, 105-108.
16. Fitzpatrick, P.F. and Massey, V. (1982) J. Biol. Chem. 257, 1166-1171.

ADSORPTION OF AROMATIC MOLECULES ON HYDROPHILIC GEL CHROMATOGRAPHY MEDIA. EFFECT OF VARIOUS SALTS

A. Sada and M. Cacace

Institute of Protein Biochemistry and Enzymology
CNR, via Toiano 6, 80072 Arco Felice, Naples, Italy

Since the time of their introduction as gel chromatography media (1), hydrophilic gels derived from cross-linked polysaccharides exhibited adsorption effects towards aromatic or apolar compounds (2,3). In view of the increasing interest towards hydrophobic interaction chromatography as a successful technique for the isolation of proteins and other biomolecules, we have undertaken a study on various parameters affecting the adsorption, such as temperature and salt concentration (4,5). This report deals with the effects that various salts, differing in their position in the Hofmeister series, exert on the elution behavior of three model compounds, ε -DNP-lysine, tryptophan and adenosine 5'-phosphate. This study is aimed at rationalizing the choice of suitable media and appropriate conditions for hydrophobic chromatography. It also represents an attempt to clarify the molecular basis of phenomena, currently poorly understood, related to hydrophobic interaction such as two-phase partition, precipitation and crystallization of proteins, and stabilization of their structures by polysaccharides and high M.W. glycols.

EXPERIMENTAL

Adenosine 5'-monophosphate (AMP), ε-dinitrophenyllysine (ε -DNP-lysine) and tryptophan were obtained from Sigma (St. Louis, MO, USA). Sephadex G-25 (Superfine, batch No. 6277) and Blue Dextran 2000 were from Pharmacia (Uppsala, Sweden). Salts were of the highest purity commercially available. When necessary, solutions were brought to neutrality by the addition of the matching acid or base. Column bed dimensions were 32 x 1 cm. Temperature was controlled with a Haake F-3C thermo-cryostat at ± 0.01°C sensitivity. Sample volume was always 200 μl and fractions of 1.1 ml were collected. Special care was taken to measure accurately column parameters. The exact position of each peak was determined graphically and the elution volume was calculated by weighing the fractions (4).

RESULTS AND DISCUSSION

The elution volume of the three model compounds from the Sephadex G-25 column was determined in the presence of various salts differing in their position in the Hofmeister series.

Figure 1 illustrates the different chromatographic behavior of the three compounds in the presence of K-phosphate buffer, $MgCl_2$ and NaSCN, respectively. Chaotropic salts, such as $MgCl_2$, cause a reduction in the adsorption on the gel for all three compounds investigated. This effect is even more evident when the chromatography is performed in the presence of NaSCN, a salt which occupies a position at one extreme in the Hofmeister series. Even in the presence of NaSCN, however, the adsorption is not abolished and each compound is eluted at a volume always larger than that corresponding to $(NH_4)_2SO_4$.

We have further investigated the effect of various salts, varying in position in the Hofmeister series, on the chromatographic behavior of the chosen aromatic compounds. Figure 2 shows the dependence of ln K_d relative to ε-DNP-lysine on salt concentration. Interestingly, the linear correlation in a wide range of concentration, demonstrated in the case of potassium phosphate (4), was found to hold for all salts. Some salts, such as NaCl and NH_4 Cl, known to exhibit poor salting-out effects, did not affect much the adsorption

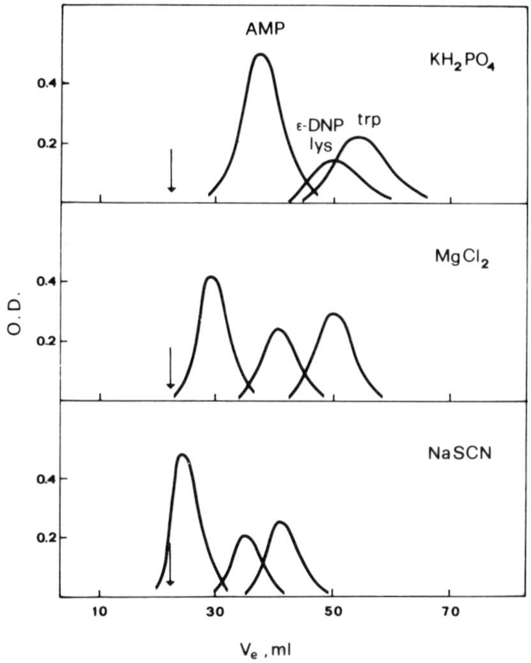

Figure 1. Elution patterns of some aromatic compounds on a Sephadex G-25 column equilibrated with salts having different positions in the Hofmeister series. Salt concentration was 1 M. The arrow indicates the elution position of $(NH_4)_2SO_4$.

of ε-DNP-lysine (as well as that of the other compounds, data not shown). Moreover, salts which exhibit good salting-in properties, such as KI or KSCN, at high concentrations tend to reduce the adsorption to the matrix.

The analogy of the process of adsorption of aromatic compounds to the hydrophilic gel to protein precipitation and salting-out is striking (6-8). It has been suggested that in hydrophobic chromatography a parameter of relevance could be the molal surface tension increment (9). However, no proportionality was observed between the latter values and the angular coefficients calculated

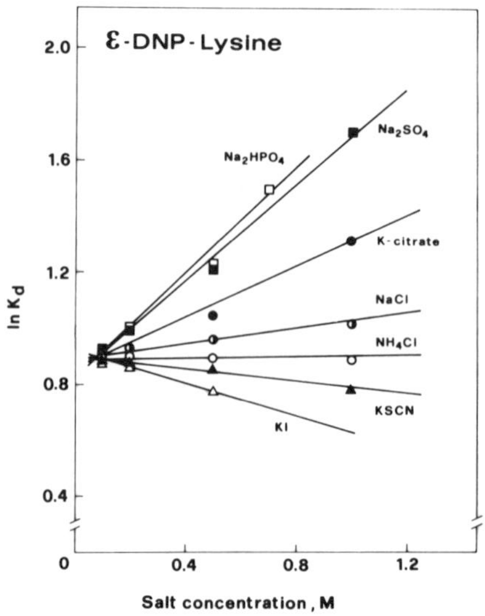

Figure 2. Dependence of ln K_d for ε-DNP-lysine on the concentration of various salts with different positions in the Hofmeister series.

from the plots in Figure 2. The linear correlation found in the wide concentration range examined for all salts indicates that the mechanism of adsorption involves the interaction of aromatic (or apolar) molecules with the hydration shell of water molecules in the matrix.

Potassium citrate and NaH_2PO_4 are water-structuring salts excluded from the coordination sphere of aromatic compounds which are therefore preferentially hydrated. This situation is favored in the presence of the water coordinated by the gel network (which is in this case of polysaccharide nature). No effect is observed in the presence of salts such as NaCl or NH_4Cl which do not exhibit strong salting-out properties. The decrease in the binding of the compounds to the column in the presence of KI or KSCN may indicate that these two latter salts also bind preferentially to ε-DNP-lysine and compete for its hydration water molecules. The persistence of an adsorption effect at the highest concentrations of such chaotropic salts is probably related to the establishment of short-range interactions with suitable groups, such as ether-linkages, present in the gel matrix (10,11). The occurrence of such interactions may also be inferred by the observation that the adsorption effect is also present at low ionic strength and that the salt dependence in all cases converges to a common limiting value for $C_{salt} \rightarrow 0$.

REFERENCES

1. Porath, J. and Flodin, P. (1959) Nature 183, 1657-1660.
2. Porath, J. (1960) Biochim. Biophys. Acta 39, 193-201.
3. Gelotte, B. (1960) J. Chromatogr. 3, 330-342.
4. Cacace, M.G. and Sada, A. (1986) J. Chromatogr. 376, 103-109.
5. Sada, A., Di Pascale, G. and Cacace, M.G. (1979) J. Chromatogr. 177, 353-356.
6. Gulewicz, K., Adamiak, D. and Sprinzl, M. (1985) FEBS Lett. 189, 179-182.
7. Porath, J. (1962) Nature 196, 47-48.
8. Arakawa, T. (1986) Arch. Biochem. Biophys. 248, 101-105.
9. Melander, W. and Horvath, C. (1977) Arch. Biochem. Biophys. 183, 200-215.
10. Determann, H. and Walter, I. (1968) Nature 219, 604-605.
11. Holmberg, L. and Linqvist, G. (1984) in: Proc. 12th Int. Carbohydrate Symp. (Vliegenthart, J.E.C. et al., eds.) p. 529, Vonk Publ., Utrecht.

Author Index

S

T

V

W

Z

Subject Index

S

T

U

V

X